Bioactive Compounds from Marine-Derived *Aspergillus, Penicillium, Talaromyces* and *Trichoderma* Species

Bioactive Compounds from Marine-Derived *Aspergillus, Penicillium, Talaromyces* and *Trichoderma* Species

Special Issue Editors

Rosario Nicoletti
Francesco Vinale

MDPI • Basel • Beijing • Wuhan • Barcelona • Belgrade

MDPI

Special Issue Editors

Rosario Nicoletti
Council for Agricultural Research and Economics
and University of Naples Federico II
Italy

Francesco Vinale
Institute for Sustainable Plant Protection, C.N.R.
and University of Naples Federico II
Italy

Editorial Office
MDPI
St. Alban-Anlage 66
4052 Basel, Switzerland

This is a reprint of articles from the Special Issue published online in the open access journal *Marine Drugs* (ISSN 1660-3397) from 2017 to 2018 (available at: https://www.mdpi.com/journal/marinedrugs/special_issues/bioactive_compounds_from_marine)

For citation purposes, cite each article independently as indicated on the article page online and as indicated below:

LastName, A.A.; LastName, B.B.; LastName, C.C. Article Title. *Journal Name* **Year**, *Article Number*, Page Range.

ISBN 978-3-03897-980-7 (Pbk)
ISBN 978-3-03897-981-4 (PDF)

Cover image courtesy of Francesco Vinale.

Contents

About the Special Issue Editors . . . . **vii**

Rosario Nicoletti and Francesco Vinale
Bioactive Compounds from Marine-Derived *Aspergillus*, *Penicillium*, *Talaromyces* and
Trichoderma Species
Reprinted from: *Mar. Drugs* **2018**, *16*, 408, doi:10.3390/md16110408 . . . . **1**

**Wensheng Li, Ping Xiong, Wenxu Zheng, Xinwei Zhu, Zhigang She, Weijia Ding and
Chunyuan Li**
Identification and Antifungal Activity of Compounds from the Mangrove Endophytic
Fungus *Aspergillus clavatus* R7
Reprinted from: *Mar. Drugs* **2017**, *15*, 259, doi:10.3390/md15080259 . . . . **4**

**Suradet Buttachon, Alice A. Ramos, Ângela Inácio, Tida Dethoup, Luís Gales, Michael Lee,
Paulo M. Costa, Artur M. S. Silva, Nazim Sekeroglu, Eduardo Rocha, Madalena M. M. Pinto,
José A. Pereira and Anake Kijjoa**
Bis-Indolyl Benzenoids, Hydroxypyrrolidine Derivatives and Other Constituents from Cultures
of the Marine Sponge-Associated Fungus *Aspergillus candidus* KUFA0062
Reprinted from: *Mar. Drugs* **2018**, *16*, 119, doi:10.3390/md16040119 . . . . **14**

**Weiyi Wang, Yanyan Liao, Chao Tang, Xiaomei Huang, Zhuhua Luo, Jianming Chen and
Peng Cai**
Cytotoxic and Antibacterial Compounds from the Coral-Derived Fungus *Aspergillus
tritici* SP2-8-1
Reprinted from: *Mar. Drugs* **2017**, *15*, 348, doi:10.3390/md15110348 . . . . **36**

**Elena V. Ivanets, Anton N. Yurchenko, Olga F. Smetanina, Anton B. Rasin,
Olesya I. Zhuravleva, Mikhail V. Pivkin, Roman S. Popov, Gunhild von Amsberg,
Shamil Sh. Afiyatullov and Sergey A. Dyshlovoy**
Asperindoles A–D and a *p*-Terphenyl Derivative from the Ascidian-Derived Fungus *Aspergillus*
sp. KMM 4676
Reprinted from: *Mar. Drugs* **2018**, *16*, 232, doi:10.3390/md16070232 . . . . **46**

**Beiye Yang, Weiguang Sun, Jianping Wang, Shuang Lin, Xiao-Nian Li, Hucheng Zhu,
Zengwei Luo, Yongbo Xue, Zhengxi Hu and Yonghui Zhang**
A New Breviane Spiroditerpenoid from the Marine-Derived Fungus *Penicillium* sp. TJ403-1
Reprinted from: *Mar. Drugs* **2018**, *16*, 110, doi:10.3390/md16040110 . . . . **58**

**Amira A. Goda, Abu Bakar Siddique, Mohamed Mohyeldin, Nehad M. Ayoub and
Khalid A. El Sayed**
The Maxi-K (BK) Channel Antagonist Penitrem A as a Novel Breast
Cancer-Targeted Therapeutic
Reprinted from: *Mar. Drugs* **2018**, *16*, 157, doi:10.3390/md16050157 . . . . **67**

**De-Sheng Liu, Xian-Guo Rong, Hui-Hui Kang, Li-Ying Ma, Mark T. Hamann and
Wei-Zhong Liu**
Raistrickiones A—E from a Highly Productive Strain of *Penicillium raistrickii* Generated through
Thermo Change
Reprinted from: *Mar. Drugs* **2018**, *16*, 213, doi:10.3390/md16060213 . . . . **88**

Rosario Nicoletti, Maria Michela Salvatore and Anna Andolfi
Secondary Metabolites of Mangrove-Associated Strains of *Talaromyces*
Reprinted from: *Mar. Drugs* **2018**, *16*, 12, doi:10.3390/md16010012 . . . . **99**

**Wenjing Wang, Xiao Wan, Junjun Liu, Jianping Wang, Hucheng Zhu, Chunmei Chen and
Yonghui Zhang**
Two New Terpenoids from *Talaromyces purpurogenus*
Reprinted from: *Mar. Drugs* **2018**, *16*, 150, doi:10.3390/md16050150 . . . . **114**

About the Special Issue Editors

Rosario Nicoletti conducts research in the field of mycology and plant pathology, with special reference to bioactive products of fungi, their role in the ecological relationships with other organisms, and perspectives for their pharmacological exploitation. He has authored over 160 scientific papers, including articles in international journals, book chapters, and communications at national and international conferences. He has served as a project reviewer and as a referee for 60 international scientific journals. He is an Editorial Board Member for the journal *Agriculture*, published by MDPI AG.

Francesco Vinale is the author of over 100 scientific papers, including articles published in scientific journals, reviews, book chapters, and abstracts and proceedings of national and international congresses. He is a reviewer for several international journals and collaborates with various companies that are involved in the development and commercialization of biopesticides and, also, methods to remediate contaminated soil and water by using microorganisms. He is responsible for or involved in numerous research projects. Research interests: (i) purification and characterization of bioactive microbial metabolites; (ii) role of microbial secondary metabolites in complex plant/antagonist/pathogen interactions; (iii) interaction between microbial antagonists, plants, and pathogens using metabolomic approaches; (iv) application of beneficial microorganisms and/or their metabolites in agriculture and industry; (v) microbial metabolites involved in pathogenic events; (vi) microbial enzymes or metabolites in decontamination of polluted soil and water (bioremediation); (vii) biochemical characterization of fungal antagonists and pathogens.

marine drugs

MDPI

Editorial

Bioactive Compounds from Marine-Derived *Aspergillus*, *Penicillium*, *Talaromyces* and *Trichoderma* Species

Rosario Nicoletti [1,*,†] **and Francesco Vinale** [2]

[1]   Council for Agricultural Research and Agricultural Economy Analysis, OFA Research Centre,
     81100 Caserta, Italy
[2]   Institute for Sustainable Plant Protection, National Research Council, 80055 Portici (NA), Italy;
     francesco.vinale@ipsp.cnr.it
*   Correspondence: rosario.nicoletti@crea.gov.it; Tel.: +39-081-253-9199
†   Current address: Department of Agriculture, University of Naples 'Federico II', 80055 Portici, Italy.

Received: 19 October 2018; Accepted: 24 October 2018; Published: 26 October 2018

The impact of bioactive compounds from natural sources on human life, particularly in pharmacology and biotechnology, has challenged the scientific community to explore new environmental contexts and the associated microbial diversity. As the largest frontier in biological discovery, the sea represents one of the most conducive reservoirs of organisms producing secondary metabolites with interesting biological activities. In the last decades fungi have received increasing attention, both for their pervasive occurrence in several habitats and for their widespread aptitude to develop symbiotic associations with higher organisms. In many cases, fungal strains have been reported as the real producers of drugs that were previously ascribed to marine plants and animals [1,2].

Species of the genera *Aspergillus*, *Penicillium*, *Talaromyces* and *Trichoderma* are renowned producers of bioactive compounds [3–5]. Until recently they were considered as 'terrestrial' fungi with merely accidental discoveries in marine environments. However, recent findings have demonstrated that actually they are very abundant in marine environments and sometimes establish symbiotic interactions with higher organisms (e.g., the case of *Aspergillus sydowii* on gorgonians [6]). It can be assumed that many species belonging to these genera of Ascomycetes are rather eclectic in their ability to adapt and thrive in very different environmental conditions. Thus, at least in terms of species number, *Aspergillus* and *Penicillium* respectively represent the first and the second most abundant genera of filamentous fungi reported from marine contexts [4,7].

Papers included in this special issue deal with marine-derived species of *Aspergillus*, *Penicillium*, *Talaromyces* and *Trichoderma*, providing a good overview of their biosynthetic potential. New compounds have been isolated and characterized from strains of *A. candidus* [8], *A. clavatus* [9], *A. tritici* [10], *P. raistrickii* [11], and *Talaromyces purpurogenus* [12]. Two papers report the recovery of strains of *Aspergillus* and *Penicillium* [13,14] that could not be ascribed to known species, thus underlying that new findings from the marine environment can expand the current taxonomic diversity and eventually contribute to a more coherent classification. Moreover, data concerning several known fungal compounds were discussed, providing clues for a better comprehension of their biosynthetic processes, and a useful indication for chemotaxonomy. Taxonomic implications and their relevance for a correct integration of new data in the current knowledge have been also discussed in a review on mangliculous strains of *Talaromyces*, after the recent separation of this genus from *Penicillium* [15].

Several novel compounds characterized from the culture filtrates of these fungi present some original or uncommon structures, such as: the raistrickiones, that represent the first case of 3,5-dihydroxy-4-methylbenzoyl derivatives of natural products [11]; the indole-diterpene alkaloids asperindoles C and D, containing a 2-hydroxyisobutyric acid residue [13]; 9,10-diolhinokiic acid, which is the first thujopsene-type sesquiterpenoid containing a 9,10-diol moiety, and roussoellol C containing a novel tetracyclic fusicoccane framework [12].

Significantly, most of the above compounds displayed biological activity as radical-scavengers [11], inhibitors of isocitrate dehydrogenase [14], antibiotic and/or cytotoxic agents. Antibiosis ranged from the antifungal activity of the coumarin, chromone, and sterone derivatives produced by *A. clavatus* R7 [9], to the antibacterial effects exhibited by several compounds towards methicillin-resistant strains of *Staphylococcus aureus*, vancomycin-resistant strains of *Enterococcus faecalis*, and *Vibrio* spp. [8,10]. The assays carried out on several human tumor cell lines were indicative of general antiproliferative effects [8,10,12–14]. In the case of penitrem A, a previously known mycotoxin, a potential for its use in cancer therapy was disclosed, based on the BK channel affinity and other side effects, which characterize this product as a possible novel sensitizing and chemotherapeutic synergizing agent [16].

In conclusion, we are grateful to all authors who contributed to our Special Issue, in the expectation that at least part of their work may have a follow up with new and exciting discoveries.

Conflicts of Interest: The authors declare no conflict of interest.

References

1. König, G.M.; Kehraus, S.; Seibert, S.F.; Abdel-Lateff, A.; Müller, D. Natural products from marine organisms and their associated microbes. *ChemBioChem* **2006**, *7*, 229–238. [CrossRef] [PubMed]
2. Thomas, T.R.A.; Kavlekar, D.P.; LokaBharathi, P.A. Marine drugs from sponge-microbe association—A review. *Mar. Drugs* **2010**, *8*, 1417–1468. [CrossRef] [PubMed]
3. Vinale, F.; Sivasithamparam, K.; Ghisalberti, E.L.; Marra, R.; Woo, S.L.; Lorito, M. *Trichoderma*–plant–pathogen interactions. *Soil Biol. Biochem.* **2008**, *40*, 1–10. [CrossRef]
4. Nicoletti, R.; Trincone, A. Bioactive compounds produced by strains of *Penicillium* and *Talaromyces* of marine origin. *Mar. Drugs* **2016**, *14*, 37. [CrossRef] [PubMed]
5. Vadlapudi, V.; Borah, N.; Yellusani, K.R.; Gade, S.; Reddy, P.; Rajamanikyam, M.; Vempati, L.N.S.; Gubbala, S.P.; Chopra, P.; Upadhyayula, S.M.; et al. *Aspergillus* secondary metabolite database, a resource to understand the secondary metabolome of *Aspergillus* genus. *Sci. Rep.* **2017**, *7*, 7325. [CrossRef] [PubMed]
6. Toledo-Hernández, C.; Zuluaga-Montero, A.; Bones-González, A.; Rodriguez, J.A.; Sabat, A.M.; Bayman, P. Fungi in healthy and diseased sea fans (*Gorgonia ventalina*): Is *Aspergillus sydowii* always the pathogen? *Coral Reefs* **2008**, *27*, 707–714. [CrossRef]
7. Jones, E.G.; Suetrong, S.; Sakayaroj, J.; Bahkali, A.H.; Abdel-Wahab, M.A.; Boekhout, T.; Pang, K.L. Classification of marine Ascomycota, Basidiomycota, Blastocladiomycota and Chytridiomycota. *Fungal Diver.* **2015**, *73*, 1–72. [CrossRef]
8. Buttachon, S.; Ramos, A.A.; Inácio, Â.; Dethoup, T.; Gales, L.; Lee, M.; Costa, P.M.; Silva, A.M.S.; Sekeroglu, N.; Rocha, E.; et al. Bis-indolyl benzenoids, hydroxypyrrolidine derivatives and other constituents from cultures of the marine sponge-associated fungus *Aspergillus candidus* KUFA0062. *Mar. Drugs* **2018**, *16*, 119. [CrossRef] [PubMed]
9. Li, W.; Xiong, P.; Zheng, W.; Zhu, X.; She, Z.; Ding, W.; Li, C. Identification and antifungal activity of compounds from the mangrove endophytic fungus *Aspergillus clavatus* R7. *Mar. Drugs* **2017**, *15*, 259. [CrossRef] [PubMed]
10. Wang, W.; Liao, Y.; Tang, C.; Huang, X.; Luo, Z.; Chen, J.; Cai, P. Cytotoxic and antibacterial compounds from the coral-derived fungus *Aspergillus tritici* SP2-8-1. *Mar. Drugs* **2017**, *15*, 348. [CrossRef] [PubMed]
11. Liu, D.S.; Rong, X.G.; Kang, H.H.; Ma, L.Y.; Hamann, M.; Liu, W.Z. Raistrickiones A−E from a highly productive strain of *Penicillium raistrickii* generated through thermo change. *Mar. Drugs* **2018**, *16*, 213. [CrossRef] [PubMed]
12. Wang, W.; Wan, X.; Liu, J.; Wang, J.; Zhu, H.; Chen, C.; Zhang, Y. Two new terpenoids from *Talaromyces purpurogenus*. *Mar. Drugs* **2018**, *16*, 150. [CrossRef] [PubMed]
13. Ivanets, E.; Yurchenko, A.; Smetanina, O.; Rasin, A.; Zhuravleva, O.; Pivkin, M.; Popov, R.S.; von Amsberg, G.; Afiyatullov, S.S.; Dyshlovoy, S. Asperindoles A–D and a *p*-terphenyl derivative from the ascidian-derived fungus *Aspergillus* sp. KMM 4676. *Mar. Drugs* **2018**, *16*, 232. [CrossRef] [PubMed]

14. Yang, B.; Sun, W.; Wang, J.; Lin, S.; Li, X.N.; Zhu, H.; Luo, Z.; Xue, Y.; Hu, Z.; Zhang, Y. A new breviane spiroditerpenoid from the marine-derived fungus *Penicillium* sp. TJ403-1. *Mar. Drugs* **2018**, *16*, 110. [CrossRef] [PubMed]
15. Nicoletti, R.; Salvatore, M.M.; Andolfi, A. Secondary metabolites of mangrove-associated strains of *Talaromyces*. *Mar. Drugs* **2018**, *16*, 12. [CrossRef] [PubMed]
16. Goda, A.A.; Siddique, A.B.; Mohyeldin, M.; Ayoub, N.M.; El Sayed, K.A. The maxi-K (BK) channel antagonist penitrem A as a novel breast cancer-targeted therapeutic. *Mar. Drugs* **2018**, *16*, 157. [CrossRef] [PubMed]

marine drugs

MDPI

Article

Identification and Antifungal Activity of Compounds from the Mangrove Endophytic Fungus *Aspergillus clavatus* R7

Wensheng Li [1], Ping Xiong [1], Wenxu Zheng [1], Xinwei Zhu [1], Zhigang She [2], Weijia Ding [1,*] and Chunyuan Li [1,*]

[1] College of Materials and Energy, South China Agricultural University, Guangzhou 510642, China; wenshengscau@126.com (W.L.); xp0000542003@scau.edu.cn (P.X.); wzheng@scau.edu.cn (W.Z.); m15521182580@163.com (X.Z.)

[2] School of Chemistry and Chemical Engineering, Sun Yat-Sen University, Guangzhou 510275, China; cesshzhg@mail.sysu.edu.cn

* Correspondence: dwjzsu@scau.edu.cn (W.D.); chunyuan-li@163.com (C.L.); Tel.: +86-20-85280319 (C.L.)

Received: 17 July 2017; Accepted: 17 August 2017; Published: 19 August 2017

Abstract: Two new coumarin derivatives, 4,4′-dimethoxy-5,5′-dimethyl-7,7′-oxydicoumarin (**1**), 7-(γ,γ-dimethylallyloxy)-5-methoxy-4-methylcoumarin (**2**), a new chromone derivative, (*S*)-5-hydroxy-2,6-dimethyl-4*H*-furo[3,4-g]benzopyran-4,8(6*H*)-dione (**5**), and a new sterone derivative, 24-hydroxylergosta-4,6,8(14),22-tetraen-3-one (**6**), along with two known bicoumarins, kotanin (**3**) and orlandin (**4**), were isolated from an endophytic fungus *Aspergillus clavatus* (collection No. R7), isolated from the root of *Myoporum bontioides* collected from Leizhou Peninsula, China. Their structures were elucidated using 1D- and 2D- NMR spectroscopy, and HRESIMS. The absolute configuration of compound **5** was determined by comparison of the experimental and calculated electronic circular dichroism (ECD) spectra. Compound **6** significantly inhibited the plant pathogenic fungi *Fusarium oxysporum*, *Colletotrichum musae* and *Penicillium italicum*, compound **5** significantly inhibited *Colletotrichum musae*, and compounds **1**, **3** and **4** greatly inhibited *Fusarium oxysporum*, showing the antifungal activities higher than those of the positive control, triadimefon.

Keywords: mangrove endophytic fungus; coumarin; chromone; sterone; antifungal activity; *Aspergillus clavatus*

---

1. Introduction

Marine mangrove endophytic fungi are among the most productive sources of structurally unusual and biologically active natural products [1–3]. *Aspergillus clavatus*, belonging to Ascomycetes (Eurotiales, Trichocomaceae), is usually found as a saprophytic fungus, which is widespread in nature, producing mycotoxins and other metabolites with activities [4–9]. In our continuous search for new bioactive natural products from mangrove endophytes, the methanol extract from the endophytic fungus, *A. clavatus* (collection No. R7) isolated from the root of *Myoporum bontioides* A. Gray collected from Leizhou Peninsula, China, had been screened to show antifungal activities against several plant pathogenic fungi [10]. This prompted us to investigate the corresponding metabolites. As a result, two new coumarin derivatives, 4,4′-dimethoxy-5,5′-dimethyl-7,7′-oxydicoumarin (**1**), 7-(γ,γ-dimethylallyloxy)-5-methoxy-4-methylcoumarin (**2**), a new chromone derivative, (*S*)-5-hydroxy-2,6-dimethyl-4*H*-furo[3,4-g]benzopyran-4,8(6*H*)-dione (**5**), and a new sterone derivative, 24-hydroxylergosta-4,6,8(14),22-tetraen-3-one (**6**), along with two known bicoumarins, kotanin (**3**) and orlandin (**4**) [11], were isolated (Figure 1). Herein, we report their isolation, structural elucidation and bioactivity.

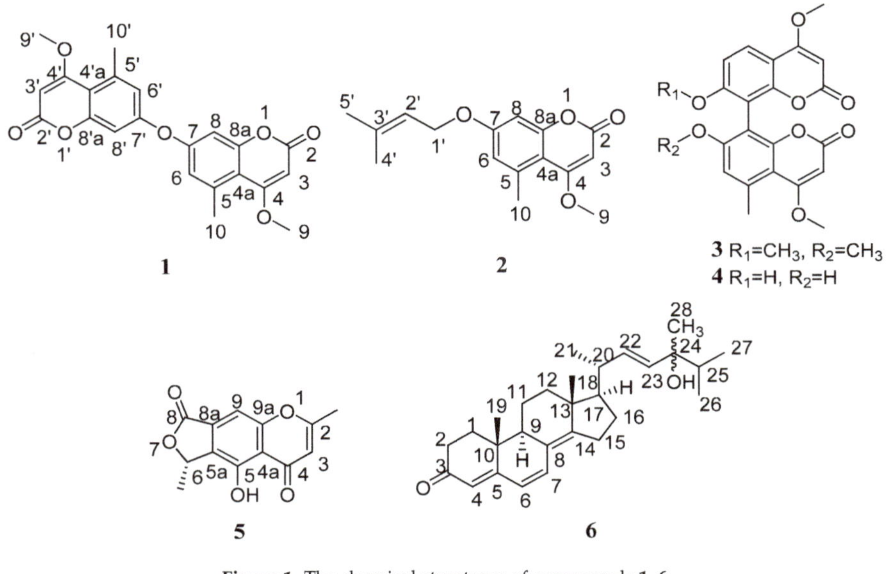

Figure 1. The chemical structures of compounds 1–6.

2. Results and Discussion

Compound **1** was obtained as a white, amorphous powder. It showed a molecular ion peak at *m/z* 395.1129 in the positive HR-ESI-MS spectrum, corresponding to molecular formula $C_{22}H_{18}O_7$ (fourteen degrees of unsaturation) ([M + H]$^+$, calcd. 395.1125). The ^{1}H NMR spectrum of **1** (Table 1) exhibited signals of two meta-coupling aromatic protons at δ_H 6.92 (d, 1H, 2.4 Hz) and 7.05 (d, 1H, 2.4 Hz), an olefinic proton at δ_H 5.70 (s, 1H), an aromatic methyl group at δ_H 2.56 (s, 3H) and a methoxyl group at δ_H 3.94 (s, 3H). The ^{13}C NMR and HSQC spectra of **1** revealed 11 carbon signals, including one methyl, one methoxyl, one ester carbonyl and eight olefinic carbons. These NMR and MS data suggested that compound **1** was most likely a symmetrical coumarin dimer derivative [12,13], wherein each subunit was substituted by one methoxyl and one methyl, and connected together by one oxygen atom. Comparison of the NMR spectral data of compound **1** with those of the known 7-hydroxy-4-methoxy-5-methylcoumarin [11] showed great similarity in that they both use deuterated dimethyl sulfoxide as solvent. However, the chemical shifts of compound **1** are obviously shifted downfield by 3.3/0.28, 6.0/0.37 ppm at C-6/H-6, C-8/H-8, and upfield by 5.7 ppm at C-7, compared with those of 7-hydroxy-4-methoxy-5-methylcoumarin, suggesting that the two coumarin subunits were presumably connected together through an oxygen atom from C-7 and C-7′ in **1**. This presumption was further confirmed by HMBC experiment (Figure 2). HMBC correlations from δ_H 6.92 (H-6/H-6′) to δ_C 109.4 (C-4a/C-4′a), 155.5 (C-7/C-7′), 105.4 (C-8/C-8′) and 23.5 (C-10/C-10′), from δ_H 2.56 (H-10/H-10′) to δ_C 109.4 (C-4a/C-4′a), 137.9 (C-5/C-5′) and 119.6 (C-6/C-6′), and from δ_H 7.05 (H-8/H-8′) to δ_C 109.4 (C-4a/C-4′a) and 156.4 (C-8a/C-8′a), suggested that the methyl (C-10/C-10′) and the oxygen atom were attached on C-5/C-5′ and C-7/C-7′, respectively. Simultaneously, HMBC correlations from δ_H 5.70 (H-3/H-3′) to δ_C 162.1 (C-2/C-2′), 169.5 (C-4/C-4′), and 109.4 (C-4a/C-4′a), and from δ_H 3.94 (H-9/H-9′) to 169.5 (C-4/C-4′), along with a four-bond HMBC correlation from δ_H 2.56 (H-10/H-10′) to δ_C 169.5 (C-4/C-4′), indicated that the methoxyl was connected to C-4/C-4′. Therefore, compound **1** was unambiguously elucidated as 4,4′-dimethoxy-5,5′-dimethyl-7,7′-oxydicoumarin.

Table 1. ^1H and ^{13}C NMR data for compounds **1** and **2**.

No.	1 [a]		2 [b]	
	δ_C	δ_H, Mult. (*J* in Hz)	δ_C	δ_H, Mult. (*J* in Hz)
1				
2	162.1, C	7.61, s	163.2, C	
2-OH				
3	88.6, CH	5.7, s	87.5, CH	5.54, s
4	169.5, C		169.8, C	
5	137.9, C		138.4, C	
6	119.6, CH	6.92, d (2.4)	116.3, CH	6.64, d (2.4)
7	155.5, C		161.2, C	
8	105.4, CH	7.05, d (2.4)	99.4, CH	6.68, d (2.4)
9	57.2, CH_3	3.94, s	55.9, CH_3	3.94, s
10	23.5, CH_3	2.56, s	23.4, CH_3	2.62, s
4a	109.4, C		107.8, C	
8a	156.4, C		156.6, C	
1'			65.1, CH_2	4.55, d (7.2)
2'	162.1, C	7.61, s	118.8, CH	5.47, t (7.2)
3'	88.6, CH	5.7, s	139.1, C	
4'	169.5, C		25.8, CH_3	1.82, s
5'	137.9, C		18.2, CH_3	1.77, s
6'	119.6, CH	6.92, d (2.4)		
7'	155.5, C			
8'	105.4, CH	7.05, d (2.4)		
9'	57.2, CH_3	3.94, s		
10'	23.5, CH_3	2.56, s		
4'a	109.4, C			
8'a	156.4, C			

[a] Measured in CD_3COCD_3; [b] Measured in $CDCl_3$.

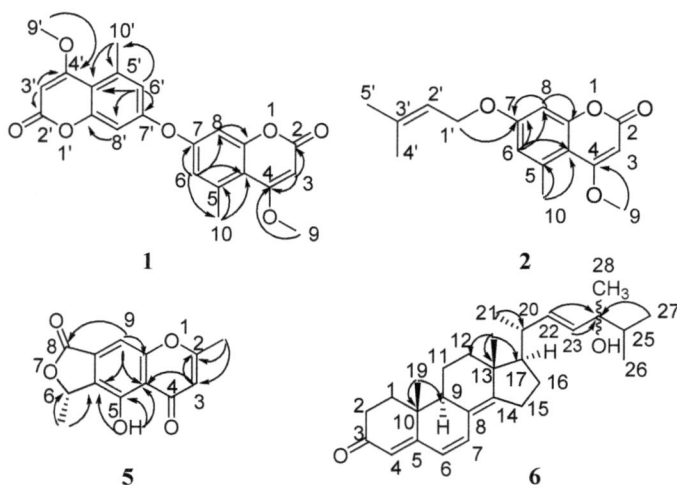

Figure 2. Selected HMBC (arrow) correlations of **1**, **2**, **5** and **6**.

Compound **2** was obtained as white needles. Its molecular formula of $C_{16}H_{18}O_4$ (eight degrees of unsaturation) was determined based on HRESIMS (*m/z* 275.1277 [M + H]$^+$, calcd. 275.1277, and 297.1106 [M + Na]$^+$, calcd. 297.1097). The ^1H NMR spectrum (Table 1) showed signals of two meta-coupling aromatic protons at δ_H 6.64 (d, 1H, 2.4 Hz) and 6.68 (d, 1H, 2.4 Hz), an olefinic proton at δ_H 5.54 (s, 1H), an aromatic methyl group at δ_H 2.62 (s, 3H), a methoxyl group at δ_H 3.94 (s, 3H), and a prenyloxy moiety at δ_H 1.77 (3H, s), 1.82 (3H, s), 5.47 (1H, t, 7.2 Hz), 4.55 (d, 2H, 7.2 Hz). The ^{13}C NMR spectrum (Table 1) exhibited 16 carbon including one methyl, one methoxyl, one ester carbonyl, one prenyl group, and eight olefinic carbons. These NMR data of **2** were similar to those of 7-hydroxy-4-methoxy-5-methylcoumarin [11]. The obvious difference between them was ascribed to a prenyl group of the former replaced the hydroxyl proton of the latter. This deduction and the position of the prenyloxy group in **2** was confirmed by comparision with the reported examples of 7-*O*-prenyl coumarins such as marianins A, B [14], and anisocoumarin B [15], and by HMBC (Figure 2) correlations from H-1′ to C-7, from H-6 to C-7, C-8, C-4a, C-10, and from H-8 to C-6, C-7, C-4a and C-8a. Additionally, the positions of the other two substituents were confirmed to be the same as 7-hydroxy-4-methoxy-5-methylcoumarin by detailed analysis of the HMBC spectrum. Thus, the structure of 4 was elucidated as 7-(*γ*,*γ*-dimethylallyloxy)-5-methoxy-4-methylcoumarin.

Compound **5** was obtained as colorless powders, and its molecular formula was established as $C_{13}H_{10}O_5$ with nine degrees of unsaturation by positive HR-ESI-MS (*m/z* 269.0423, [M + Na]$^+$, calcd. 269.0420). The characteristic UV absorption maxima at 229, 242, 263, 345 nm suggested the presence of a chromone pattern in **5** [16,17]. The ^1H and ^{13}C NMR spectral data of **5** are listed in Table 2. The ^1H NMR spectrum exhibited signals of one olefinic methyl at δ_H 2.52 (s, 1H), one secondary methyl at δ_H 1.67 (d, 6.6 Hz 3H) connected to one oxomethine at δ_H 5.73 (q, 6.6 Hz, 1H), one hydroxyl at δ_H 13.43 (s, 1H), and two aromatic proton singlets at δ_H 6.37 and 7.83. The olefinic methyl was revealed to be attached at C-2 due to HMBC correlations (Figure 2) from the 2-CH$_3$ proton at δ_H 2.52 to C-2 (δ_C 170.3) and C-3 (δ_C 108.9), and from the aromatic H-3 proton (δ_H 6.37) to C-2 and C-4a (δ_C 112.7). The hydroxyl was proved to be substituted at C-5 based on HMBC correlations from 5-OH (δ_H 13.43) to C-4a, C-5 (δ_C 155.8) and C-5a (δ_C 130.7). These results, combined with the HMBC correlations, including H-9 (δ_H 7.37) to C-4a, C-5a, and the oxygen-bearing C-9a (δ_C 157.2), ambiguously established the chromone substructure, indicating that the positions of C-8a and C-9 of the aromatic ring were substituted by the remaining moiety. Subsequently, HMBC correlations from H-9 to C-8 (δ_C 168.2), from H-6 (δ_H 5.73) to C-5a, C-8, 6-CH$_3$ (δ_H 1.67), from 6-CH$_3$ to C-5a, C-6 (δ_C 76.5), together with the remaining 2 degrees of unsaturation revealed by the molecular formula, suggested a *γ*-valerolactone ring system attached to C-8a and C-9 through C-8 and C-6, respectively. Thus, the planar structure of **5** was completely established. The absolute configuration of **5** was determined by comparing the theoretical calculation of ECD (electronic circular dichroism) with the experimental ECD [18,19]. The experimental ECD of **5** is similar to the ECD of the (*S*)-model compound (Figure 3), so as to determine that the absolute configuration of **5** was 6*S*. Therefore, the structure of **5** was as shown in Figure 1.

Table 2. ^1H and ^{13}C NMR data for compound **5** in CD$_3$COCD$_3$.

No.	δ_C	δ_H, Mult. (*J* in Hz)
1		
2	170.3, C	
2-CH$_3$	19.7, CH$_3$	2.52, s
3	108.9, CH	6.37, s
4	184.0, C	
4a	112.7, C	
5	155.8, C	
5-OH		13.43, s
5a	130.7, C	
6	76.5, CH	5.73, q (6.6)
6-CH$_3$	18.2, CH$_3$	1.67, d (6.6)
8	168.2, C	
8a	131.0, C	
9	102.9, CH	7.37, s
9a	157.2, C	

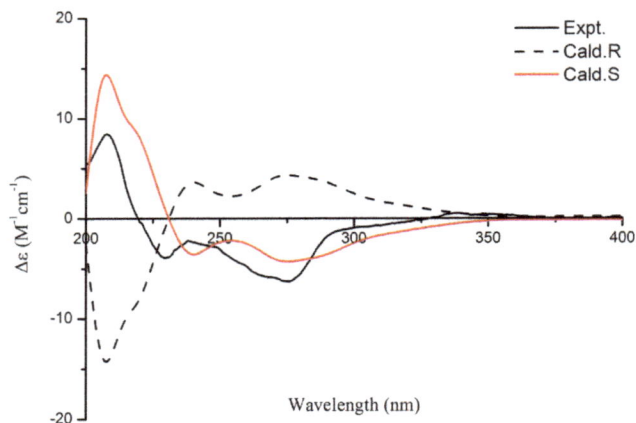

Figure 3. The calculated and experimental ECD spectra of **5**.

Compound **6** was obtained as colorless needles. The molecular formula was determined as $C_{28}H_{40}O_2$ (nine degrees of unsaturation) by analysis of positive HR-ESI-MS (*m/z* 409.3108; [M + H]$^+$, calcd. 409.3101). The ^{1}H NMR spectrum of **6** displayed five olefinic proton signals at δ_H 6.04 (d, 1H, 9.6 Hz), 6.61 (d, 1H, 9.6 Hz), 5.75 (s, 1H), 5.48 (m, 1H), 5.49 (m, 1H), six methyl signals at δ_H 0.98 (s, 3H), 1.00 (s, 3H), 1.08 (d, 3H, 6.6 Hz), 0.91 (d, 3H, 3.1 Hz), 0.90 (d, 3H, 3.2 Hz), 1.23 (s, 3H), and numerous methene and methine signals ranging from δ_H 1.29 to 2.53. The ^{13}C NMR and HSQC spectra showed 28 carbons, including a ketone group (δ_C 199.5), and four olefinic double bonds. Comparison of the ^{1}H and ^{13}C NMR spectral data (Table 3) of compound **6** with those of ergosta-4,6,8(14),22-tetraen-3-one [20], revealed their great structural similarities. However, the ^{13}C NMR of the former exhibited one quaternary carbon more at δ_C 74.9 and one methine less at high field than in the latter. This result combined the difference between their molecular formulas presumed **6** to be a hydroxyl substituted derivate of ergosta-4,6,8(14),22-tetraen-3-one. The position of the hydroxyl group was confirmed to locate at C-24 by HMBC correlations of H-22, H-23, H-26, H-27 and H-28 to C-24 at δ_C 74.9. Detailed analysis of HSQC, ^{1}H-^{1}H-COSY, and HMBC spectra (Figure 2) allowed the complete assignment of the proton and carbon signals of **6**. The relative configuration of **6** was assigned by NOESY (nuclear overhauser effect spectroscopy) experiments. In the NOESY spectrum of **6**, NOE (nuclear overhauser effect) correlations of Me-18 with both H-20 and H-11a, and the lack of NOE correlations between Me-18 and H-11b, suggested β-orientations of Me-18, H-20 and H-11a. Consequently, NOE correlations between H-19 and H-11a, and the absence of NOE correlations between H-19 and H-11b suggested H-19 was also in a β-orientation. Additionally, NOE correlations between H-9 and H-1a, and between H-19 and H-1b, along with no NOE correlations between H-19 and H-1a, indicated H-9 was in an α-orientation. The configuration of double bond Δ22 was deduced to be *E* by comparison of the chemical shifts with those of the same positions of ergosta-4,6,8(14),22-tetraen-3-one and the large coupling constant (15.2 Hz) between H-22 and H-23. The configuration of C-24 could not be assigned based on the obtained NOE data. Therefore, compound **6** was elucidated as 24-hydroxylergosta-4,6,8(14),22-tetraen-3-one, as shown in Figure 1.

Table 3. ^{1}H and ^{13}C NMR data for compound **6** in CDCl$_3$.

No.	δ_C	δ_H, Mult. (*J* in Hz)
1	34.1, CH$_2$	a1.82, m b2.02, m
2	34.1, CH$_2$	a2.46, m b2.53, m
3	199.5, C	
4	123.0, CH	5.75, s
5	124.5, C	
6	124.6, CH	6.04, d (9.6)
7	134.0, CH	6.61, d (9.6)
8	164.3, C	
9	44.3, CH	2.14, m
10	36.7, C	
11	19.0, CH$_2$	a1.60, m b1.71, m
12	35.6, CH$_2$	a1.31, m b2.09, m
13	44.0, C	
14	155.7, C	
15	25.2, CH$_2$	a2.39, m b2.48, m
16	27.8, CH$_2$	a1.49, m b1.80, m
7	55.9, CH	1.29, m
18	19.0, CH$_3$	0.98, s
19	16.6, CH$_3$	1.00, s
20	39.1, CH	2.22, m
21	21.0, CH$_3$	1.08, d (6.6)
22	133.7, CH)	5.48, dd (8.2, 15.2)
23	134.0, CH	5.52, d (15.2)
24	74.9, C	
25	38.1, CH	1.70, m
26	17.6, CH$_3$	0.91, d (3.1)
27	17.2, CH$_3$	0.90, d (3.2)
28	25.4, CH$_3$	1.23, s

In addition, the structures of the known compounds **3** and **4** [11] were identified by comparison of their spectroscopic data with those reported in the literature. HRESIMS, ^{1}H, ^{13}C, ^{1}H-^{1}H COSY, HSQC and HMBC NMR spectra of the new compounds are available at the Supplementary Materials File (Figures S1–S22).

The antifungal activities of the isolated compounds were examined in vitro towards three plant pathogens, including *Fusarium oxysporum* Schlecht. f. sp. lycopersici (Sacc.) W.C. Snyder et H.N. Hansen (*F. oxysporum*), *Colletotrichum musae* (Berk. and M. A. Curtis) Arx. (*C. musae*), and *Penicillium italicum* Wehme (*P. italicm*). From the results presented in Table 4, all of the compounds showed broad-spectrum inhibitory activities against these fungi except compound **2**, which is inactive towards *P. italicm* with MIC value >729.66 μM. Moreover, compound **6** exhibited the strongest broad-spectrum inhibitory activities against all the three pathogenic fungi *F. oxysporum*, *C. musae* and *P. italicm* with MIC values of 244.73, 195.79 and 61.18 μM, respectively, in comparison with other compounds and triadimefon (used as the positive control, MIC values = 340.43, 272.39, 170.24 μM, respectively). In addition, compounds **1**, **3** and **4** showed high activities against *F. oxysporum* (MICs = 253.81, 235.85, 252.47 μM, respectively), which was better than triadimefon. Whereas compound **5** displayed more potent inhibitory activity against *C. musae*, with MIC values of 203.07 μM, than triadimefon.

Table 4. Antifungal activity of the isolated compounds by MIC values (μM).

Compounds	*F. oxysporum*	*C. musae*	*P. italicm*
1	253.81	380.71	253.81
2	729.66	547.25	>729.66
3	235.85	353.77	235.85
4	252.47	378.71	252.47
5	609.21	203.07	304.61
6	244.73	195.79	61.18
Triadimefon [a]	340.43	272.39	170.24

[a] positive control.

3. Experimental Section

3.1. General Experimental Procetures

Melting points were determined using a JH30 melting point detector (Jia Hang Instrument Co., Ltd., Shanghai, China). Optical rotations were measured using a Horiba SEPA-300 polarimeter at 25 °C. The UV spectra were obtained on a Shimadzu UV-2550 spectrophotometer (Shimadzu, Tokyo, Japan), and IR spectra were run on a Nicolet 5DX-Fourier transform infrared spectrophotometer (Thermo Electron Corporation, Madison, WI, USA). NMR spectra data were recorded at Bruker AV-600 MHz NMR spectrometers (Bruker Biospin AG, Fällanden, Switzerland), with tetramethylsilane (TMS) as internal standard, and the chemical shifts were reported in δ values (ppm). The HRESIMS spectra were recorded on an Q-TOF mass spectrometer (Thermo Fisher, Frankfurt, Germany). CD spectra were recorded with a Chirascan™ CD spectrometer (Applied Photophysics, Leatherhead, UK). Silica gel (200–300 mesh) for column chromatography was purchased from Qingdao Haiyang Chemical Co., Ltd., Qingdao, China. Sephadex LH-20 was purchased from Amersham Pharmacia Biotech. Buckinghamshire, UK. All other chemicals were of analytical grade.

3.2. Fungal Material and Fermentation

The fungal strain R7 was isolated from the root of *M. bontioides*, collected from the mangrove in Leizhou peninsula, China, in May 2014, and deposited at the College of Materials and Energy, South China Agricultural University, Guangdong Province, China. The strain has been identified as *A. clavatus*, according to morphologic traits and molecular identification [10]. Its 599 base pair ITS sequence had 99% sequence identity to those of several *A. clavatus* strains (AY373847.1, NR121482.1, KF669481.1) by a NCBI BLAST search. The sequence data has been submitted to GenBank with accession number KY765893.

A small agar scrap with mycelium of the fungal isolate which was grown on potato dextrose agar medium for 5 days at 28 °C was added into 250 mL GYT medium (1% glucose, 0.1% yeast extract, 0.2% peptone, 0.2% crude sea salt), and incubated at 28 °C, 180 rpm for 6 days as seed culture. Then the seed culture was grown on a solid autoclaved rice substrate medium (one hundred 1000 mL Erlenmeyer flasks, each containing 100 mL water, 100 g rice and 0.3 g crude sea salt) for 30 days at 25 °C under static stations.

3.3. Extraction and Isolation

The mycelia and solid rice medium were extracted with 95% ethanol three times. The solvent was concentrated to 1 L *in vacuo* and extracted with equal volume of ethyl acetate, yielding 70.0 g extract. Then the extract was subjected to a silica gel column (30 × 6 cm), eluting with gradient of petroleum ether/ethyl acetate (97:3, 95:5, 75:25, 50:50, 25:75, 0:100, *v/v*) to afford six fractions (Fr. A1–Fr. A6). Fraction A2 was chromatographed on Sephadex LH-20 CC (110 × 4 cm) eluting with Methanol-dichloromethane-petroleum ether (2:2:1, *v/v*), to obtain three subfractions (Fr. A2-1–Fr.

A2-3) based on TLC properties. Fraction A2-3 was dissolved in acetone and recrystallized at room temperature to afford compound **5** (8.2 mg). Fraction A3 was purified by preparative silica gel TLC (petroleum ether/ethyl acetate, 5:1, *v/v*) to yield compound **6** (15 mg). Fraction A4 was further fractioned by silica gel eluting with petroleum ether-ethyl acetate (85:15, 75:25, 50:50, *v/v*) to give three subfractions (Fr. A4-1–Fr. A4-3). Fraction A4-1 was separated through Sephadex LH-20 CC (methanol-dichloromethane 3:2, *v/v*) to afford compound **2** (7.5 mg). Fraction A4-3 was applied to preparative silica gel TLC (petroleum ether/ethyl acetate, 1:5, *v/v*) to give compounds **3** (6.8 mg) and **4** (4.3 mg). Fraction A6 was subjected to silica gel column chromatography and eluted with ethyl acetate/methanol (50:50, 15:85, 0:100, *v/v*), leading to three subfractions (Fr. A6-1–Fr. A6-3). Fraction A6-3 was further chromatographed on a Sephadex LH-20 column using methanol/dichloromethane (3:2, *v/v*) to afford compound **1** (1.8 mg).

4,4′-dimethoxy-5,5′-dimethyl-7,7′-oxydicoumarin (**1**): White amorphous powder. m.p. 174.7–175.3 °C; HR-ESI-MS *m/z* 315.1129 ([M + H]$^+$, calcd. for $C_{22}H_{19}O_7$ 315.1125). ^{1}H NMR and ^{13}C NMR data see Table 1.

7-(γ,γ-dimethylallyloxy)-5-methoxy-4-methylcoumarin (**2**): White crystal. m.p. 115.5–116.3 °C; UV (EtOH) λ_{max} (log ε): 208 (4.28), 218 (4.08), 310 (3.81) nm; IR (KBr) ν_{max}: 3144, 2968, 1716, 1613, 1575, 1400, 1256, 1152 cm^{-1}; HR-ESI-MS *m/z* 275.1127 ([M + H]$^+$, calcd. for $C_{16}H_{19}O_4$ 275.1127). ^{1}H NMR and ^{13}C NMR see Table 1.

(S)-5-hydroxy-2,6-dimethyl-4H-furo[3,4-g]benzopyran-4,8(6H)-dione (**5**): White needles. $[\alpha]_D^{25}$ = −37.87 (c 0.0015, MeOH); UV (MeOH) λ_{max} (log ε): 229 (4.32), 242 (4.24), 263 (3.72), 345 (3.63) nm; IR (KBr) ν_{max}: 3420, 2987, 1635, 1616, 1487, 1396, 1173 cm^{-1}; HR-ESI-MS *m/z* 269.0423 ([M + Na]$^+$, calcd. for $C_{13}H_{10}O_5Na$ 269.0420). ^{1}H NMR and ^{13}C NMR see Table 2.

24-hydroxylergosta-4,6,8(14),22-tetraen-3-one (**6**): Yellow oil. $[\alpha]_D^{25}$ = +173.3 (c 0.004, MeOH); UV (MeOH) λ_{max} (log ε): 341 (3.79) nm; IR (KBr) ν_{max}: 3420, 3136, 1669, 1650, 1528, 1453, 1401, 1385 cm^{-1}; HR-ESI-MS *m/z* 409.3108 ([M + H]$^+$, calcd. for $C_{28}H_{41}O_2$ 409.3101). ^{1}H NMR and ^{13}C NMR see Table 3.

3.4. Computational Analyses

Conformational analyses for compound **5** were performed via Spartan'10 software (Wavefunction, Inc., Irvine, CA, USA) using the MMFF94 molecular mechanics force field calculation. Conformers within a 10 kcal/mol energy window were generated and optimized using DFT calculations at the B3LYP/6-31G (d) level. Conformers for R or S were chosen for ECD calculations in MeOH at the B3LYP/6-311+G (2d, p) level. Rotary strengths for a total of 50 excited states were calculated. The IEF-PCM solvent model for MeOH was used. The calculated ECD spectra were obtained by density functional theory (DFT) and time-dependent DFT (TD-DFT) using Gaussian 09 (Gaussian Inc., Wallingford, CT, USA) program package. The calculated ECD curve was generated using SpecDis 1.6 software package (University of Wurzburg, Wurzburg, Germany) with a half-bandwidth of 0.2 eV.

3.5. Antifungal Activity Assay

The following four phytopathogenic fungi were used for bioassay: *F. oxysporum, C. musae,* and *P. italicm.* They were obtained from the College of Agriculture, South China Agricultural University. The antifungal activities of the isolated compounds were determined by the broth dilution method as described in the previous report to get the minimum inhibitory concentration (MIC) [21]. Triadimefon and the solvent were used as positive and negative control, respectively.

4. Conclusions

In conclusion, two new coumarin derivatives, 4,4′-dimethoxy-5,5′-dimethyl-7,7′-oxydicoumarin (**1**), 7-(γ,γ-dimethylallyloxy)-5-methoxy-4-methylcoumarin (**2**), a new chromone derivative, (S)-5-hydroxy-2,6-dimethyl-4*H*-furo[3,4-g]benzopyran-4,8(6*H*)-dione(**5**), and a new sterone derivative,

24-hydroxylergosta-4,6,8(14),22-tetraen-3-one (**6**), together with two known bicoumarins, kotanin (**3**) and orlandin (**4**), were isolated from an endophytic fungus *Aspergillus clavatus* R7, isolated from the root of *Myoporum bontioides* that collected from Leizhou Peninsula, China. Compound **6** remarkably inhibited *Fusarium oxysporum, Colletotrichum musae* and *Penicillium italicum*, compound **5** highly inhibited *Colletotrichum musae*, and compounds **1, 3** and **4** greatly inhibited *Fusarium oxysporum*, by comparison to triadimefon, indicating that these compounds could be used as leads of new fungicides.

Supplementary Materials: The NMR and MS spectra of **1, 2, 5** and **6** are available online at www.mdpi.com/link/1660-3397/15/8/259/s1 in Figures S1–S22.

Acknowledgments: This work was supported by the National Natural Science Foundation of China (21102049), the Natural Science Foundation of Guangdong Province (2015A030313405, 9451064201003751), the Science and Technology Project for public welfare research and capacity building of Guangdong Province (2016A020222019), the Science and Technology Project of Guangzhou City (201707010342), and the Scientific Research Foundation for the Returned Overseas Chinese Scholars, State Education Ministry (grant number [2015] 311).

Author Contributions: C.L. and W.D. conceived and designed the experiments; W.L., P.X., W.Z., X.Z. and W.D. performed the experiments; Z.S. and C.L. analyzed the data; W.L. and W.D. wrote the paper. C.L. revised and edited the manuscript.

Conflicts of Interest: The authors declare no conflict of interest.

References

1. Blunt, J.W.; Copp, B.R.; Keyzers, R.A.; Munroa, M.H.G.; Prinsep, M.R. Marine natural products. *Nat. Prod. Rep.* **2017**, *34*, 235–294. [CrossRef] [PubMed]
2. Ahmed, A.M.M.; Taha, T.M.; Abo-Dahab, N.F.; Hassan, F.S.M. Process optimization of L-glutaminase production; a tumour inhibitor from marine endophytic isolate *Aspergillus* sp. ALAA-2000. *J. Microb. Biochem. Technol.* **2016**, *8*, 256–267. [CrossRef]
3. Gao, S.; Li, X.; Williams, K.; Proksch, P.; Ji, N.; Wang, B. Rhizovarins A–F, indole-diterpenes from the mangrove-derived endophytic fungus *Mucor irregularis* QEN-189. *J. Nat. Prod.* **2016**, *79*, 2066–2074. [CrossRef] [PubMed]
4. Wang, J.; Huang, Y.; Fang, M.; Zhang, Y.; Zheng, Z.; Zhao, Y.; Su, W. Brefeldin A, a cytotoxin produced by *Paecilomyces* sp. and *Aspergillus clavatus* isolated from *Taxus mairei* and *Torreya grandis*. *FEMS Immunol. Med. Microbiol.* **2002**, *34*, 51–57. [CrossRef] [PubMed]
5. Bawin, T.; Seye, F.; Boukraa, S.; Zimmer, J.; Raharimalala, F.N.; Ndiaye, M.; Compere, P.; Delvigne, F.; Francis, F. Histopathological effects of *Aspergillus clavatus* (Ascomycota: Trichocomaceae.) on larvae of the southern house mosquito, *Culex quinquefasciatus* (Diptera: Culicidae). *Fungal Biol.* **2016**, *120*, 489–499. [CrossRef] [PubMed]
6. Zhang, C.; Zheng, B.; Lao, J.; Mao, L.; Chen, S.; Kubicek, C.P.; Lin, F. Clavatol and patulin formation as the antagonistic principle of *Aspergillus clavatonanicus*, an endophytic fungus of *Taxus mairei*. *Appl. Microbiol. Biotechnol.* **2008**, *78*, 833–840. [CrossRef] [PubMed]
7. Losada, L.; Ajayi, O.; Frisvad, J.C.; Yu, J.J.; Nierman, W.C. Effect of competition on the production and activity of secondary metabolites in *Aspergillus* species. *Med. Mycol.* **2009**, *47*, S88–S96. [CrossRef] [PubMed]
8. Wang, J.; Bai, G.; Liu, Y.; Wang, H.; Li, Y.; Yin, W.; Wang, Y.; Lu, F. Cytotoxic metabolites produced by the endophytic fungus *Aspergillus clavatus*. *Chem. Lett.* **2015**, *44*, 1148–1149. [CrossRef]
9. Liu, J.Y.; Song, Y.C.; Zhang, Z.; Wang, L.; Guo, Z.J.; Zou, W.X.; Tan, R.X. *Aspergillus fumigatus* CY018, an endophytic fungus in *Cynodon dactylon* as a versatile producer of new and bioactive metabolites. *J. Biotechnol.* **2004**, *114*, 279–287. [CrossRef] [PubMed]
10. Ding, W.; Zhang, S.; Gong, B.; Li, C.; Wang, X. Isolation and inhibitory activity of endophytic fungi from the semi-mangrove plant *Myoporum bontioides* A. Gray. *Guangdong Agric. Sci.* **2014**, *41*, 74–78.
11. Hüttel, W.; Müller, M. Regio- and stereoselective Intermolecular oxidative phenol coupling in kotanin biosynthesis by *Aspergillus niger*. *ChemBioChem* **2007**, *8*, 521–529. [CrossRef] [PubMed]
12. Ju-ichi, M.; Takemura, Y.; Okano, M.; Fukamiya, N.; Hatano, K.; Asakawa, Y.; Hashimoto, T.; Ito, C.; Furukawa, H. The structures of claudimerines-A and -B, novel bicoumarins from *Citrus hassaku*. *Chem. Pharm. Bull.* **1996**, *44*, 11–14. [CrossRef]

13. Mahibalan, S.; Rao, P.C.; Khan, R.; Basha, A.; Siddareddy, R.; Masubuti, H.; Fujimoto, Y.; Begum, A.S. Cytotoxic constituents of *Oldenlandia umbellata* and isolation of a new symmetrical coumarin dimer. *Med. Chem. Res.* **2016**, *25*, 466–472. [CrossRef]

14. Fukuda, T.; Sudoh, Y.; Tsuchiya, Y.; Okuda, T.; Fujimori, F.; Igarashi, Y. Marianins A and B, prenylated phenylpropanoids from *Mariannaea camptospora*. *J. Nat. Prod.* **2011**, *74*, 1327–1330. [CrossRef] [PubMed]

15. Ngadjui, T.B.; Ayafor, J.F.; Sondengam, B.L.; Connolly, J.D. Coumarins from *Clausena anisata*. *Phytochemistry* **1989**, *28*, 585–589. [CrossRef]

16. Huang, M.; Li, J.; Liu, L.; Yin, S.; Wang, J.; Lin, Y. Phomopsichin A–D; four new chromone derivatives from mangrove endophytic fungus *Phomopsis* sp. 33#. *Mar. Drugs* **2016**, *14*, 215. [CrossRef]

17. Xia, M.; Cui, C.; Li, C.; Wu, C.; Peng, J.; Li, D. Rare chromones from a fungal mutant of the marine-derived *Penicillium purpurogenum* G59. *Mar. Drugs* **2015**, *13*, 5219–5236. [CrossRef] [PubMed]

18. Srebro-Hooper, M.; Autschbach, J. Calculating natural optical activity of molecules from first principles. *Annu. Rev. Phys. Chem.* **2017**, *68*, 399–420. [CrossRef] [PubMed]

19. Pescitelli, G.; Bruhn, T. Good Computational Practice in the Assignment of Absolute Configurations by TDDFT Calculations of ECD Spectra. *Chirality* **2016**, *28*, 466–474. [CrossRef] [PubMed]

20. Fujimoto, H.; Nakamura, E.; Okuyama, E.; Ishibashi, M. Six immunosuppressive features from an ascomycete, *Zopfiella longicaudata*, found in a screening study monitored by immunomodulatory activity. *Chem. Pharm. Bull.* **2004**, *52*, 1005–1008. [CrossRef] [PubMed]

21. Wang, J.; Ding, W.; Wang, R.; Du, Y.; Liu, H.; Kong, X.; Li, C. Identification and bioactivity of compounds from the mangrove endophytic fungus *Alternaria* sp. *Mar. Drugs* **2015**, *13*, 4492–4504. [CrossRef] [PubMed]

marine drugs

MDPI

Article

Bis-Indolyl Benzenoids, Hydroxypyrrolidine Derivatives and Other Constituents from Cultures of the Marine Sponge-Associated Fungus *Aspergillus candidus* KUFA0062

Suradet Buttachon [1,2], Alice A. Ramos [1,2], Ângela Inácio [1,2], Tida Dethoup [3], Luís Gales [1,4], Michael Lee [5], Paulo M. Costa [1,2], Artur M. S. Silva [6], Nazim Sekeroglu [7], Eduardo Rocha [1,2], Madalena M. M. Pinto [2,8], José A. Pereira [1,2,*] and Anake Kijjoa [1,2,*]

[1] ICBAS-Instituto de Ciências Biomédicas Abel Salazar, Rua de Jorge Viterbo Ferreira, 228, 4050-313 Porto, Portugal; lgales@ibmc.up.pt (L.G.); pmcosta@icbas.up.pt (P.M.C.); erocha@icbas.up.pt (E.R.)

[2] Interdisciplinary Centre of Marine and Environmental Research (CIIMAR), Terminal de Cruzeiros do Porto de Lexões, Av. General Norton de Matos s/n, 4450-208 Matosinhos, Portugal; nokrari_209@hotmail.com (S.B.); ramosalic@gmail.com (A.A.R.); angelainacio@gmail.com (Â.I.)

[3] Department of Plant Pathology, Faculty of Agriculture, Kasetsart University, Bangkok 10240, Thailand; tdethoup@yahoo.com

[4] Instituto de Biologia Molecular e Celular (i3S-IBMC), Universidade do Porto, Rua de Jorge Viterbo Ferreira, 228, 4050-313 Porto, Portugal

[5] Department of Chemistry, University of Leicester, University Road, Leicester LE 7 RH, UK; ml34@leicester.ac.uk

[6] Departamento de Química & QOPNA, Universidade de Aveiro, 3810-193 Aveiro, Portugal; artur.silva@ua.pt

[7] Medicinal and Aromatic Plant Programme, Plant and Animal Sciences Department, Vocational School, Kilis 7 Aralık University, 79000 Kilis, Turkey; nsekeroglu@gmail.com

[8] Laboratório de Química Orgânica, Departamento de Ciências Químicas, Faculdade de Farmácia, Universidade do Porto, Rua de Jorge Viterbo Ferreira, 228, 4050-3 13 Porto, Portugal; madalena@ff.up.pt

* Correspondence: jpereira@icbas.up.pt (J.A.P.); ankijjoa@icbas.up.pt (A.K.); Tel.: +351-22-042-8331 (J.A.P. & A.K.); Fax: +351-22-206-2232 (J.A.P. & A.K.)

Received: 14 March 2018; Accepted: 5 April 2018; Published: 6 April 2018

Abstract: A previously unreported *bis*-indolyl benzenoid, candidusin D (**2e**) and a new hydroxypyrrolidine alkaloid, preussin C (**5b**) were isolated together with fourteen previously described compounds: palmitic acid, clionasterol, ergosterol 5,8-endoperoxides, chrysophanic acid (**1a**), emodin (**1b**), six *bis*-indolyl benzenoids including asterriquinol D dimethyl ether (**2a**), petromurin C (**2b**), kumbicin B (**2c**), kumbicin A (**2d**), 2″-oxoasterriquinol D methyl ether (**3**), kumbicin D (**4**), the hydroxypyrrolidine alkaloid preussin (**5a**), (3S, 6S)-3,6-dibenzylpiperazine-2,5-dione (**6**) and 4-(acetylamino) benzoic acid (**7**), from the cultures of the marine sponge-associated fungus *Aspergillus candidus* KUFA 0062. Compounds **1a**, **2a–e**, **3**, **4**, **5a–b**, and **6** were tested for their antibacterial activity against Gram-positive and Gram-negative reference and multidrug-resistant strains isolated from the environment. Only **5a** exhibited an inhibitory effect against *S. aureus* ATCC 29213 and *E. faecalis* ATCC29212 as well as both methicillin-resistant *S. aureus* (MRSA) and vancomycin-resistant enterococci (VRE) strains. Both **1a** and **5a** also reduced significant biofilm formation in *E. coli* ATCC 25922. Moreover, **2b** and **5a** revealed a synergistic effect with oxacillin against MRSA *S. aureus* 66/1 while **5a** exhibited a strong synergistic effect with the antibiotic colistin against *E. coli* 1410/1. Compound **1a**, **2a–e**, **3**, **4**, **5a–b**, and **6** were also tested, together with the crude extract, for cytotoxic effect against eight cancer cell lines: HepG2, HT29, HCT116, A549, A 375, MCF-7, U-251, and T98G. Except for **1a**, **2a**, **2d**, **4**, and **6**, all the compounds showed cytotoxicity against all the cancer cell lines tested.

Keywords: *Aspergillus candidus*; Aspergillaceae; sponge-associated fungus; *bis*-indolyl benzenoids; hydroxypyrrolidine; antibacterial activity; cytotoxicity

1. Introduction

Aspergillus candidus (Family Aspergillaceae) is a member of *Aspergillus* section *Candidi* [1]. This species frequently contaminates stored food and feeding stuff [2], and is one of the most frequently encountered mold in cereal grains and flour [3]. It also occurs in soil, usually on seeds or in the rhizosphere, and also in milk [4]. Strains of *A. candidus* produce a variety of secondary metabolites including chlorflavonin, a chlorine containing flavone antifungal antibiotic [5], *p*-terphenyl derivatives such as terphenyllin [6], deoxyterphenyllin [7], 3-hydroxyterphenyllin [8], candidusins A and B [9], immunosuppressant terprenins [10]. On the other hand, there are only a few reports on the chemical investigation of marine-derived *A. candidus*. Wei et al. [11] reported the isolation of cytotoxic prenylterphyllin, 4"-deoxyprenylterphyllin, 4"-deoxyisoterprenin and 4"-deoxyterprenin from *A. candidus* IF10, isolated from marine sediment. Recently, Wang et al. [12] described the isolation of spiculisporic acid derivatives including the new compounds, spiculisporic acids F and G.

In our pursuit for antibiotic and anticancer compounds from marine-derived fungi from the tropical sea, we have investigated secondary metabolites from cultures of *A. candidus* KUFA 0062, which were isolated from the marine sponge *Epipolasis* sp., collected from the coral reef at the Similan Island National Park in Phang-Nga province, Southern Thailand.

Chromatographic fractionation and further purification of the crude ethyl acetate extract of the cultures of *A. candidus* KUFA 0062, furnished two previously undescribed compounds named candidusin D (**2e**) and preussin C (**5b**), as well as the previously reported chrysophanic acid (**1a**) [13], emodin (**1b**) [14], six *bis*-indolyl benzenoids including asterriquinol D dimethyl ether (**2a**) [15], petromurin C (**2b**) [16], kumbicin B (**2c**) [15], kumbicin A (**2d**) [15], 2"-oxoasterriquinol D methyl ether (**3**) [17], kumbicin D (**4**) [15], the hydroxypyrrolidine alkaloid preussin (**5a**) [18–20], (3*S*, 6*S*)-3,6-dibenzylpiperazine-2,5-dione (**6**) [21], and 4-(acetylamino) benzoic acid (**7**) [22] (Figure 1). Additionally, the common fungal metabolites, i.e., palmitic acid, clionasterol [23], and ergosterol 5,8-endoperoxides [14] were also isolated (Supplementary Material, Figure S1).

Compounds **1a**, **2a–e**, **3**, **4**, **5a–b**, and **6** were tested for their antibacterial activity against four reference bacterial strains consisting of two Gram-positive (*Staphylococcus aureus* ATCC 29213 and *Enterococcus faecalis* ATCC 29212) and two Gram-negative bacteria (*Escherichia coli* ATCC 25922 and *Pseudomonas aeruginosa* ATCC 27853), three multidrug-resistant isolates from the environment, MRSA *S. aureus* 66/1, VRE *E. faecalis* B3/101, a colistin-resistant *E. coli* 1418/1, and a clinical isolate ESBL *E. coli* SA/2. The isolated compounds were also investigated for their capacity to inhibit biofilm formation in the four reference strains as well as for their potential synergism with the clinically used antibiotics against multidrug-resistant isolates from the environment. Moreover, these compounds were also evaluated for their cytotoxic effect against eight cancer cell lines, i.e., Hep G2 (human hepatocellular carcinoma), HT29 (human colorectal adenocarcinoma), HCT116 (human colorectal carcinoma), A549 (human lung carcinoma), A375 (human malignant melanoma), MCF7 (human mammary gland adenocarcinoma), U251 (human glioblastoma multiforme), and T98G (human glioblastoma astrocytoma), by MTT assay.

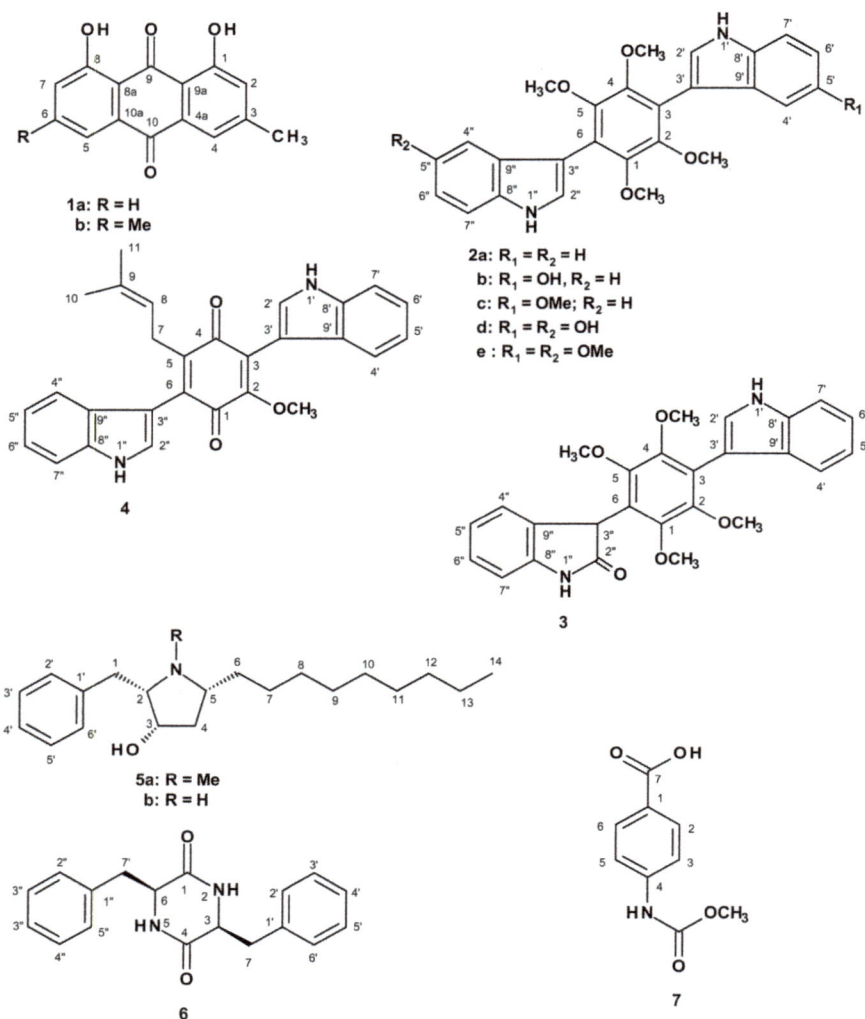

Figure 1. Structures of some secondary metabolites isolated from the cultures of the marine sponge-associated fungus *A. candidus* KUFA 0062.

2. Results and Discussion

The structures of palmitic acid, clionasterol [22], ergosterol 5,8-endoperoxides [14], chrysophanic acid (**1a**) [13], emodin (**1b**) [14], asterriquinol D dimethyl ether (**2a**) [15], petromurin C (**2b**) [16], kumbicin B (**2c**) [15], kumbicin A (**2d**) [15], 2″-oxoasterriquinol D methyl ether (**3**) [17], kumbicin D (**4**) [15], preussin (**5a**) [18–20], (3*S*, 6*S*)-3,6-dibenzylpiperazine-2,5-dione (**6**) [21] and 4-(acetylamino) benzoic acid (**7**) [22] were elucidated by analysis of their 1D and 2D NMR spectra as well as HRMS data, and also by comparison of their spectral data to those reported in the literature (Supplementary Materials, Figures S2–S17, S20–S25, S32–S35 and Tables S1–S4).

Compound **2e** was isolated as a white solid (m.p. 299–300 °C), and its molecular formula $C_{28}H_{28}N_2O_6$ was established based on its (+)-HRESIMS *m/z* 489.2030 [M + H]+, (calculated

489.2026 for $C_{28}H_{29}N_2O_6$), indicating sixteen degrees of unsaturation. The IR spectrum showed absorption bands for amine (3346 cm^{-1}), aromatic (1625, 1579 cm^{-1}), and ether (1291 cm^{-1}) groups. The general features of the ^{1}H and ^{13}C NMR spectra of **2e** (Table 1, Supplementary Materials, Figures S18 and S19) resembled those of kumbicin A (**2d**), petromurin D (**2b**), and two *bis*-indolyl benzenoids isolated from the soil fungus *Aspergillus kumbius* [15]. The ^{13}C NMR spectrum (Table 1) exhibited twelve carbon signals which, in combination with DEPTs and HSQC spectra, can be categorized in six quaternary sp^2 (δ_C 153.1, 147.6, 131.0, 127.4, 122.0, 106.7), four methine sp^2 (δ_C 126.0, 111.9, 110.9, 102.1), and two methoxyl (δ_C 60.3, 55.3) carbon signals. As the number of the carbon signals is less than the number of carbon atoms found in HRMS, the molecule must be symmetrical. The presence of the 3,5-disubstituted indole moiety with the methoxyl group on C-5 was supported by the presence of an amine proton at δ_H 11.15 (d, *J* = 2.2 Hz), a doublet at δ_H 7.43 (*J* = 2.5 Hz, H-2; δ_C 126.0), and the proton signals of the 1,2,3 substituted benzene ring at δ_H 6.78, dd (*J* = 8.8, 2.5 Hz, H-6; δ_C 110.9), 6.88, d (*J* = 2.5 Hz, H-4; δ_C 102.1), 7.33, dd (*J* = 8.8, H-7; δ_C 111.9) as well as a singlet of OMe at δ_H 3.72 (δ_C 55.3). That another portion of the molecule was 2,3,5,6-tetramethoxy-1,6-disubstituted benzene ring was supported by the presence of the carbon signals at δ_H 122.0 (C-1 and C-6) and δ_C 147.6 (C-2, C-3, C-5, C-6) as well as four chemically and magnetically equivalent methoxyl groups at δ_H 3.46 (δ_C 60.3). Putting together the HRMS and NMR data, the structure of **2e** was established as 5'-methoxykumbicin B. However, instead of having one methoxyl group and one hydroxyl group on C-5' and C-5" of the indole ring system like petromurin D, the substituents on both C-5' and C-5" are methoxyl groups in **2e**. To the best of our knowledge, **2e** is a new analogue of *bis*-indolyl benzenoids, which was named candidusin D.

Table 1. ^{1}H and ^{13}C NMR (DMSO, 300.13 and 75.4 MHz) and HMBC assignment for **2e**.

Position	δ_C, Type	δ_H, (*J* in Hz)
1, 2, 4, 5	147.6, C	-
3, 6	120.0, C	-
NH-1', 1"	-	11.15, d (2.2)
2', 2"	126.0, CH	7.43, d (2.5)
3', 3"	106.7, C	-
4', 4"	102.1, CH	6.88, d (2.5)
5', 5"	153.1, C	-
6', 6"	110.9, CH	6.78, dd (8.8, 2.5)
7', 7"	111.9, CH	7.33, d (8.8)
8', 8"	131.0, C	-
9', 9"	127.7	-
OCH$_3$-1, 2, 4, 5	60.3, CH$_3$	3.46, s
OCH$_3$-5, 5'	55.3, CH2	3.72, s

Compound **3** was isolated as white crystals (m.p. 290–292 °C) with $[\alpha]_D^{25}$ − 115.4 (*c* 0.04, MeOH). Analysis of the (+)-HRESIMS, ^{1}H, ^{13}C NMR, COSY, HSQC, HMBC data of **3** (Table S3, Supplementary Materials, Figures S20 and S21), revealed that this compound has the same planar structure as that 2"-oxoasterriquinol D methyl ether, an oxidized *bis*-indolyl benzenoid previously isolated from extracts of the sclerotia of *Aspergillus sclerotiorum* (NRRL 5167) [17]. Although Whyte et al. [17] have found that 2"-oxoasterriquinol D methyl ether was optically active, with $[\alpha]_D^{25}$ − 13.9 (*c* 0.001 g/mL, MeOH), they did not determine the absolute configuration of the stereogenic carbon (C-3") of the oxidized indole ring system of this compound. In order to fully elucidate the structure of **3**, we have attempted to obtain suitable crystal for X-ray analysis. The results obtained, as seen in the ORTEP (Oak Ridge Thermal-Ellipsoid Plot Program) view in Figure 2, revealed that although **3** was levorotatory, it was not optically pure since both enantiomers were present in the crystal.

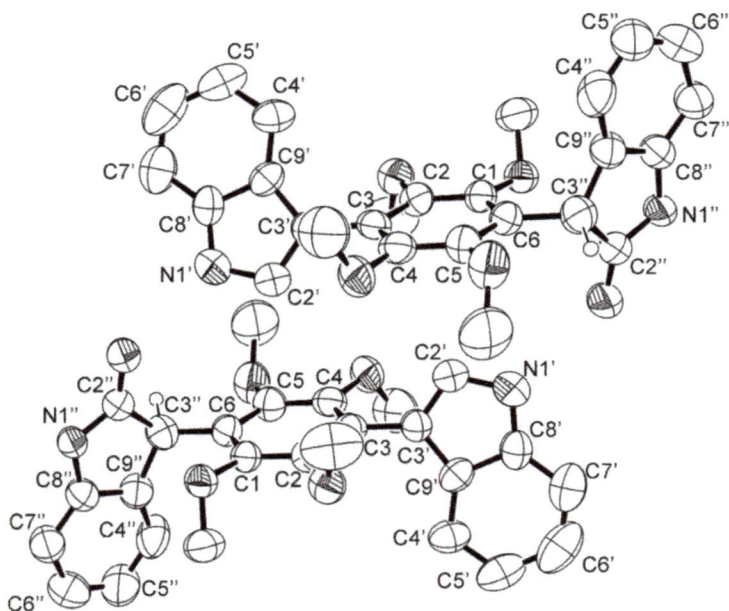

Figure 2. ORTEP (Oak Ridge Thermal-Ellipsoid Plot Program) view of **3**.

The (+)-HRESIMS, ^{1}H, ^{13}C NMR (Table 2), COSY, HSQC, and HMBC (Supplementary Materials, Figures S24 and S25) data of **5a** revealed that its planar structure is equal to that of the antifungal compound L-657, 398, first isolated from the fermentation of *Aspergillus ochraceus* ATCC 22947 [18]. However, the authors failed to establish the relative stereochemistry of this compound by ^{1}H NMR data. The same compound, called preussin, was later isolated as a yellow oil ($[\alpha]_D^{25}$ + 22.0, *c* 1.0, CHCl$_3$), from the mycelial cake obtained from fermentation of a fungus *Preussia* sp. [19]. By comparison of the chemical shift values of H$_2$-1 and H$_2$-6 of the (*S*)-and (*R*)-*O*-methylmandelate esters with those of the acetate derivative of preussin, Johnson et al. [19] determined the absolute configurations of C-2, C-3, and C-5 of preussin as 2*S*, 3*S*, and 5*R*.

Since we were able to obtain **5a** as a suitable crystal, X-ray analysis was carried out to determine if the absolute configurations of its stereogenic carbons were the same as those of preussin. The ORTEP view of a protonated **5a**, shown in Figure 3, not only determined the number of carbon atoms in the alkyl side chain, but also confirmed the absolute configurations of C-2, C-3, and C-5 as 2*S*, 3*S*, and 5*R*. To the best of our knowledge, this is the first X-ray crystal structure of preussin.

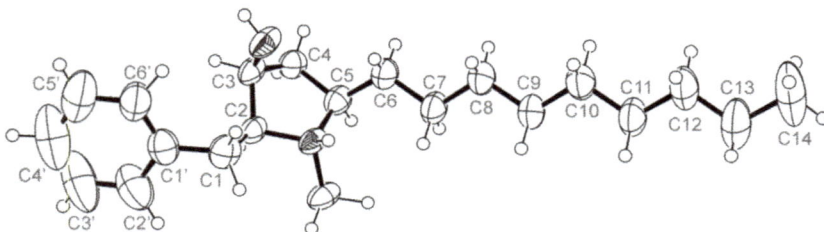

Figure 3. ORTEP view of a protonated **5a**.

Compound **5b** was isolated as a white crystal (m.p. 219–220 °C), and its molecular formula was established as $C_{20}H_{33}NO$ on the basis of its (+)-HRESIMS *m/z* 304.2647 [M + H]$^+$, (calculated 304.2640 for $C_{20}H_{34}NO$), which indicated five degrees of unsaturation. The IR spectrum showed absorption bands for hydroxyl and amine (3441 cm^{-1}) and aromatic (1634, 1581, 1539 cm^{-1}). The general features of the ^{1}H and ^{13}C NMR spectra of **5b** resembled those of **5a** (Table 2, Supplementary Materials, Figures S26, S27 and S29), except for the absence of the *N*-methyl singlet at δ_H 2.81 (δ_C 36.7), which was confirmed by its molecular formula. Moreover, H$_2$-1, H-3, H-4, and H$_2$-6 appeared at lower frequencies while H-2 and H-5 resonated at higher frequencies when compared to the corresponding protons in **5a**. Additionally, the chemical shift values of C-2 and C-5 in **5b** are around 10 ppm lower than those of the corresponding carbons in **5a**, while C-6 in **5b** is around 3 ppm higher than that of C-6 in **5a**. Therefore, **5b** is a *N*-demethyl analog of **5a**. This hypothesis was confirmed by COSY correlations from H-2 (δ_H 3.46, m; δ_C 65.2) to H-1a (δ_H 3.11, dd, *J* = 13.8, 8.0 Hz; δ_C 31.3), H-1b (δ_H 2.94, dd, *J* = 13.8, 6.6 Hz; δ_C 31.3) and H-3 (δ_H 4.09, dd, *J* = 7.9, 4.0 Hz; δ_C 68.7); H-3 to H-2, H-4β (δ_H 2.37, ddd, *J* = 13.9, 9.9, 5.5 Hz; δ_C 38.5) and OH-3 (δ_H 6.62, d, *J* = 4.0 Hz); H-5 (δ_H 4.09, dd, *J* = 7.9, 4.0 Hz; δ_C 68.7) to H-4β, H-4α (δ_H 1.58, dd, *J* = 13.9, 5.8 Hz; δ_C 38.5) and H$_2$-6 (δ_H 1.73, m/1.64, m; δ_C 33.6) as well as from the methyl triplet at δ_H 0.86, t, (*J* = 6.8 Hz) to a brs at δ_H 1.26 (Table 2 and Supplementary Materials, Figures S28 and 29). This was supported by HMBC correlations from H-1a and H-1b to C-2 and C-3; H-2 to C-1; H-3 to C-2, C-5; OH-3 to C-2, C-3, C-4; H-4 to C-3, C-5 and C-6, and from H$_2$-6 to C-4, C-5, C-7 (Table 2 and Supplementary Materials, Figure S30). The relative stereochemistry of the pyrrolidine ring of **5b** was established, based on NOESY experiments, to be the same as that of **5a**. The NOESY spectrum (Table 2, Supplementary Materials, Figure S31) exhibited correlations from OH-3 to H-1a, H-3 and H-4α; H-3 to H-2, OH-3 and H-4β; H-4β to H-3, H-4α and H-5, and from H-6 at δ_H 1.73, m to H-4α. Since **5a** can be hypothesized to derive from **5b** through *N*-methylation by SAM, it is legitimate to presume that the absolute configurations of C-2, C-3, and C-5 in **5b** should be the same as those of **5a**, i.e., 2*S*, 3*S*, 5*R*.

Since **5b** could not be obtained as a suitable crystal for X-ray analysis, we have performed a calculated Electronic Circular Dichroism (ECD) spectrum to compare with its experimental ECD spectrum to determine unequivocally the absolute configurations of its C-2, C-3, and C-5. Experimental NOESY data of **5b**, as well as the data obtained from X-ray crystallographic and Trost's *O*-methylmandelate method [19] of **5a**, suggest that all the pyrrolidine substituents of **5b** are directed to the same side of the ring. This formally narrows the investigation of the stereochemistry of **5b** to the enantiomer pair (2*S*, 3*S*, 5*R*) and (2*R*, 3*R*, 5*S*), combined with the two possible configurations of the secondary amine. However, comparison of ECD simulated spectra for the above-mentioned configurations of **5b**, in its neutral forms, with the experimental spectrum in methanol was inconsistent with the crystallographic configuration previously determined for its N-CH$_3$ analogue (**5a**). Since this analogue, most probably a metabolic derivative of **5b** and therefore sharing the same (2*S*, 3*S*, 5*R*) configuration, precipitated and crystalized as an ammonium salt, the same was assumed for **5b** and ECD simulations were therefore redirected to its protonated (positively charged) form.

This was expected to be a major difference since the calculated SCF HOMO and LUMO orbitals show that the electronic transitions happen between the two rings. The new charge in the five-membered ring alters significantly the electron density distribution and the molecular orbitals energies. The electronic transitions associated with the hydrocarbon side-chain, however, do not involve the five-membered ring and should remain unaffected. Additionally, these transitions are predicted to be caused by wavelengths below 200 nm, hence outside the observational window.

Table 2. ¹H and ¹³C NMR of **5a** (DMSO, 300.13 and 75.4 MHz) and **5b** (DMSO, 300.13 and 75.4 MHz) and HMBC and NOESY assignments for **5b**.

Position	5a (DMSO, 300.13 and 75.4 MHz)		5b (DMSO, 500.13 and 125.4 MHz)				
	δ_C, Type	δ_H, (J in Hz)	δ_C, Type	δ_H, (J in Hz)	COSY	HMBC	NOESY
1a	30.6, CH₂	3.69, dd (13.0, 9.4)	31.3, CH₂	3.11, dd (13.8, 8.0)	H-1b, 2	C-2,3	H-1b, 2′, 6′
b		3.13, dd (12.7, 4.8)		2.94, dd (13.8, 6.6)	H-1a, 2	C-2,3	H-1a
2	76.0, CH	3.09, q (4.7)	65.5, CH	3.46, m	H-1b, 1b, 3	C-1	H-3
3	68.6, CH	4.22, m	68.7, CH	4.09, dd (7.9, 4.0)	H-2, 4β, OH-3	C-2,5	H-2, H-4β, OH-3
4α	38.8, CH₂	2.06, ddd (14.8, 9.3, 2.1)	38.5, CH₂	1.58, dd (13.9, 5.8)	H-3, 5	C-3,5,6	H-3, 4β, 5
β		2.69, ddd (15.0, 8.2, 8.2)		2.37, ddd (13.9, 9.9, 5.5)	H-5	C-2,3,5,6	H-4α, H-3
5	69.3, CH	2.92, m	57.6, CH	3.41, m	H-4β, 4α, 6	C-4,6,7	H-4α
6	30.1, CH₂	2.25, m	33.6, CH₂	1.73, m	H-5	C-4,5,7	
		1.95, m		1.64, m	H-5	C-4,5,7	
7	26.5, CH₂	1.40, m	26.1, CH₂	1.26, brs			
8	26.5, CH₂	1.22, brs	28.6, CH₂	1.26, brs			
9	29.4, CH₂	1.22, brs	28.9, CH₂	1.26, brs			
10	29.2, CH₂	1.22, brs	28.8, CH₂	1.26, brs			
11	29.2, CH₂	1.22, brs	28.7, CH₂	1.26, brs			
12	29.4, CH₂	1.22, brs	31.6, CH₂	1.26, brs			
13	22.7, CH₂	1.22, brs	22.1, CH₂	1.26, brs			
14	14.1, CH₃	0.88, t (6.5)	14.0, CH₃	0.86, t (6.8)	H-13	C-12,13	
1′	136.1, C	-	137.4, C	-			
2′	129.4, CH	7.31, m	129.1, CH	7.31, m		C-1	H-1a, 1b, H2
3′	128.9, CH	7.35, m	128.5, CH	7.34, m		C-1	
4′	127.2, CH	7.25, m	126.6, CH	7.26, m		C-2′, 6′	
5′	128.9, CH	7.31, m	128.5, CH	7.34, m		C-1	
6′	129.4, CH	7.35, m	129.1, CH	7.31, m		C-1	H-1a, 1b, H2
N-CH₃	36.7, CH₃	2.81, s	-				
OH-3	-	5.12, d (11.7)	-	5.62, d (4.0)	H-3	C-2,3,4	H-1a, 3, 4β

Models were constructed based only on the protonated (2*S*, 3*S*, 5*R*) configuration because deprotonation upon dissolution of the secondary ammonium group is expected to be negligible. The typically large pKa values (≈10) associated with these very weak acid groups in water is maintained in methanol [24]. To have a good representation of **5b**'s conformational space, the three staggered orientations of C-1 (relative to C-2) and of 3-OH (relative to C-3) were combined with two five-membered ring conformations to give 18 models. One, the lowest MM2 energy, staggered conformation of C-6 was set in all models, as well as a staggered linear conformation for the carbon chain. The orientation of the phenyl group was always determined as a result of energy minimization. To have a notion of its energetic ranking, the conformational energy of each of the 18 models was first minimized using the semi-empirical PM6 method. Since the six lowest PM6-energy conformations accounted for 98% of the conformer population, as determined by a Boltzmann distribution of computed PM6 conformational energies, these were further minimized with the Amplitude Probability Density Function (APFD) Density Functional Theory (DFT) method. The four lowest-energy APFD conformers were selected for ECD simulation, also by the Boltzmann distribution significance. Figure 4 shows the lowest energy model. The ECD transitions of the four models were calculated to a minimum wavelength of 170 nm with the TDDFT method at the same level of theory than the final minimizations (APFD, with the same basis set and solvent model). The four sets of transitions were Boltzmann weighed using the calculated populations, Gaussian broadened by 0.17 eV and added to give the averaged simulated spectrum in Figure 5. As is apparent, the fit with the experimental spectrum is good, representing well its three main negative intensities (roughly at 215, 225 and 255 nm). This evidence supports the claim that the isolated **5b** is the (2*S*, 3*S*, 5*R*) enantiomer (Figure 4) in the ammonium form.

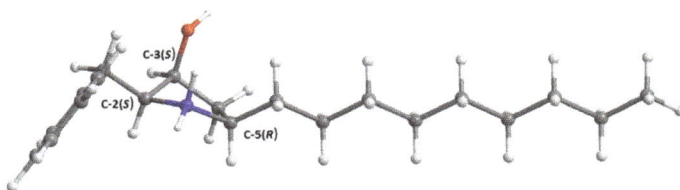

Figure 4. The minimal APDF (Amplitude Probability Density Function) DFT (Density Functional Theory) energy molecular model of amino-protonated **5b**, in its (2*S*, 3*S*, 5*R*) configuration. The other 17 models tested differ in the conformation of the ring and in the orientations of the hydroxyl and -CH$_2$-C$_6$H$_6$ substituent groups.

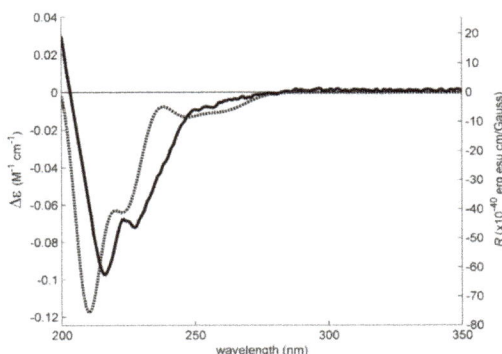

Figure 5. Experimental (solid line) and simulated (dotted line) methanol ECD (Electronic Circular Dichroism) spectra of **5b**. The simulated spectrum of the (2*S*, 3*S*, 5*R*) model configuration fits well the experimental data.

Chrysophanic acid (**1a**), asterriquinol D dimethyl ether (**2a**), petromurin C (**2b**), kumbicin B (**2c**), kumbicin A (**2d**), 2"-oxoesterriquinol D methyl ether (**3**), kumbicin D (**4**), preussin (**5a**), preussin C (**5b**), and (3*S*, 6*S*)-3,6-dibenzylpiperazine-2,5-dione (**6**) were tested for their antibacterial activity and their minimum inhibitory concentration (MIC) and minimum bactericidal concentration (MBC) for four reference strains consisting of three multidrug-resistant isolates from the environment and one clinical isolate. In the range of concentrations tested, none of the compounds were active against Gram-negative bacteria. However, **5a** displayed an inhibitory effect against Gram-positive bacteria, with MIC values of 32 µg/mL for *S. aureus* ATCC 29213 and *E. faecalis* ATCC 29212. Moreover, **5a** consistently showed a MIC value of 32 µg/mL for both methicillin-resistant *S. aureus* (MRSA) and vancomycin-resistant enterococci (VRE) strains. Compound **5a** also displayed a MBC value of 64 µg/mL for the VRE *E. faecalis* B3/101 strain, however, it was not possible to determine the MBC for the other Gram-positive strains. This result suggests that **5a** might not only have a bacteriostatic effect but also a bactericidal action.

The effect of **1a**, **2a–e**, **3**, **4**, **5a**, **b**, and **6** on biofilm formation in four reference bacterial strains was also evaluated. Four concentrations of **5a**, ranging from 2 × MIC to $\frac{1}{4}$ MIC were tested against *S. aureus* ATCC 2913 and *E. faecalis* ATCC 2912. Since it was not possible to determine a MIC of the rest of the other compounds, the previously tested highest concentration that did not inhibit bacterial growth was used. Although none of the compounds tested showed an inhibitory effect on biomass production in *P. aeruginosa* ATCC 27853, the anthraquinone chrysophanic acid (**1a**) induced significant reduction in biofilm formation (67.7 ± 8.3% of control; One-sample *t* test: * *p* < 0.05, significantly different from 100%) in *E. coli* ATCC 25922. Interestingly, **5a** also caused an almost 50% reduction of biofilm mass production (42.8 ± 32.7% of control) at a sub-inhibitory concentration in *E. coli* ATCC 25922. Although this inhibitory effect was not statistically significant, this result shows an interesting aspect in that **5a** is active against both Gram-positive and Gram-negative bacteria. In fact, among the compounds tested, only **5a** was capable of interfering with biofilm formation in *S. aureus* ATCC 29213 and *E. faecalis* ATCC 29212 at concentrations equal to or above the MIC (Figure 6), however, at lower concentrations, **5a** had no effect on the biomass production of both strains.

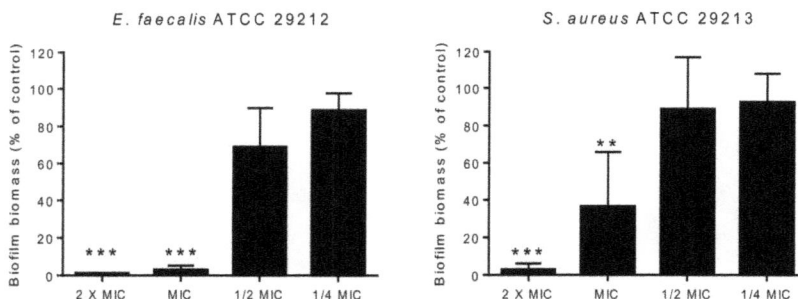

Figure 6. Gram-positive bacteria biofilm biomass production after 24 h of incubation with different concentrations of compound 5a. Data are shown as Mean ± SD of the three independent experiments. One-sample t test: ** *p* < 0.01 and *** *p* < 0.001, significantly different from 100%. MIC, minimum inhibitory concentration.

The tested compounds were also screened for their potential synergies with the clinically relevant antibiotics and some of them revealed small to moderate synergistic associations with antibiotic as determined by the disk diffusion method. Except **5a**, which presented an inhibition zone of 8 mm for VRE *E. faecalis* B3/101 and MRSA *S. aureus* 66/1, the rest of the tested compounds showed no inhibition zone (0 mm), when tested alone. The combination of **2a–d**, **3**, **4**, **5a**, and **6** with cefotaxime in impregnated disks resulted in a minor synergistic effect, as can be observed by a small increase in the zone of inhibition when compared with the halo of inhibition produced by cefotaxime alone in an ESBL *E. coli* strain (SA/2). In the case of VRE *E. faecalis* B3/101, all the tested compounds induced a small increase in the halo of a partial inhibition of vancomycin when compared with vancomycin alone, however, this effect was more pronounced with **5a**. On the other hand, only **2b** and **5a** revealed a synergistic effect with oxacillin against MRSA *S. aureus* 66/1. Interestingly, although **2b** alone did not exhibit the antibacterial effect, it showed a marked synergistic effect with oxacillin, increasing the halo of inhibition from 0 mm, when oxacillin was used alone, to 7 mm when tested in combination. These results were also confirmed by the checkerboard method since the combination of **5a** with vancomycin and oxacillin showed the values of ΣFIC ≤ 0.5 against VRE and MRSA isolates, respectively (Table 3).

Table 3. Fractional inhibitory concentration (FIC) index of **5a** in combination with clinically relevant antibiotics obtained by the checkerboard method.

Bacterial Strain	5a-Van		5a-Ox	
	ΣFIC	Activity	ΣFIC	Activity
E. faecalis B3/101 (VRE)	0.4	S	-	-
S. aureus 66/1 (MRSA)	-	-	0.2	S

S = synergism; VAN = vancomycin; OX = oxacillin.

Interestingly, **5a** also demonstrated a strong synergistic effect with colistin, reducing MIC of colistin from 8 μg/mL to less than 0.008 μg/mL, i.e., at least 100-fold, when tested against *E. coli* 1410/1 (Table 4). Following this result, lower concentrations of **5a** were tested and it was found that a concentration of 8 μg/mL of **5a** was enough to alter *E. coli* 1410/1 resistance to colistin, thus decreasing colistin MIC to 1 μg/mL (susceptibility breakpoint of ≤0.2 μg/mL) [25].

Table 4. Combined effect of colistin with **5a** against colistin resistant *E. coli* strain 1410/1. MICs for colistin are expressed in μg/mL.

Compound	μg/mL of 5a + Colistin							
5a	0	1	2	4	8	16	32	64
Colistin (MIC)	8 [R]	8 [R]	8 [R]	4 [R]	1 [S]	0.016 [S]	<0.008 [S]	<0.008 [S]

MIC = minimum inhibitory concentration; [R] = resistant; [S] = sensitive.

With the exception of **2b**, the modest synergistic effects observed by the disk diffusion method were not replicated when the MIC of the antibiotics was determined in the presence of the tested compounds. The MIC of oxacillin, in a combination with **2b**, was four folds less than that of oxacillin alone (from 64 μg/mL to 16 μg/mL) when tested against MRSA *S. aureus* 66/1. Although no effect on MIC of the antibiotics was observed with any other compounds, it is likely that since the method used for determination of the MIC requires two-fold serial dilutions, the synergistic effect between the antibiotics and the test compounds might not be enough to decrease the MIC of the antibiotics by a factor of two, in case the absolute difference between the two consecutive concentrations is large. This reasoning can justify the difference in the results obtained using two different methods.

The cytotoxic effect of the ethyl acetate crude extract of *A. candidus* KUFA 0062 (200 μg/mL) and **1a**, **2a–e**, **3**, **4**, **5a**, **b**, and **6**, at only one concentration (100 μM), were tested against eight cancer cell lines, i.e., Hep G2, HT29, HCT116, A549, A375, MCF7, U251, and T98G by MTT assay. At 200 μg/mL, the crude extract significantly decreased cell viability in all cell lines tested, with higher cytotoxic effect in A375 > HT29 > HCT116 > HepG2 > A549 > T98G > MCF7 > U251 cells. At 100 μM, **5a** was found to decrease the cell viability in all cell lines tested, reaching the lowest % of cell viability in colon cancer cell lines (6.4% and 8.6% in HCT116 and HT29, respectively). Compound **5b** also decreased cell viability in all of the cell lines tested, however, it was less effective than **5a**. Except for human malignant glioma U251 and T98G cells, **2b** significantly decreased the number of viable cells in all the cell lines tested (with cell viability less than 50). Similarly, **2e** and **3** also exhibited a significant decrease of cell viability in all cell lines tested except for T98G and HepG2 for **2e** and HepG2 for **3**. Compound **2c** significantly decreased cell viability in the HepG2, HT29, HCT116, A549, and A375 cells. Compounds **1a**, **2a**, **2d**, **4** and **6**, at 100 µM, did not change cell viability in any of the cancer cell lines tested (Table 5). In view of these results, dose-response curves were constructed to determine IC_{50} values only for compounds and cell lines whose significant cytotoxic effects were observed. Doxorubicin (Dox) was used as a positive control. The IC_{50} values and respective 95% confidence intervals are summarized in Table 6. Compound **5a** was the most effective in all the cell lines tested with IC_{50} ranging from 12.3 µM in HT29 cells to 74.1 µM in U251 cells while **2b** also induced a significant decrease of cell viability with IC_{50} values ranging from 34.8 µM in H29 cells to 94.8 µM in MCF7 cells. On the other hand, **3** was more cytotoxic in HT29 cells with IC_{50} value of 43.2 µM but less cytotoxic in U251 cells with IC_{50} value of 120.2 µM. Compound **5b** showed IC_{50} values ranging from 57.2 µM (in HT29 cells) to 215.7 µM (in A549 cells). Compound **2c** and **2e** significantly decreased cell viability but with higher IC_{50} values ranging from 72.9 µM (**2c** in HCT116 cells) to 186.6 µM (**2e** in MCF7 cells). Over all, it can be observed that while **2b**, **3**, **5a**, and **5b** were more effective in HT29 cells, **2c** and **2e** were more active in HCT116 cells. This suggests that these compounds may act by different mechanisms of action since each cell line has different genetic characteristics. As to the positive control, Dox, we have observed a decrease in cell viability in all cell lines with IC_{50} values ranging from 0.05 µM (in A375 cells) to 15.4 µM (in T98G cells). Our findings are in agreement with the literature showing that HT29 was more resistant to the cytotoxic effect of Dox than the HCT116 cells [26].

Table 5. Percentage of cell viability relative to control of **1a**, **2a–e**, **3**, **4**, **5a**, **b**, **6** (100 μM) and crude extract of *A. candidus* KUFA 0062 (200 μg/mL) in eight human cancer cell lines after 48 h of incubation as assessed by MTT assay.

Compounds/Extract	Percentage of Cell Viability, Mean (SD)							
	HepG2	HT29	HCT116	A549	A375	MCF7	U251	T98g
control	100 (8.7)	100 (11.3)	100 (12.8)	100 (8.5)	100 (6.6)	100 (12.3)	100 (19.4)	100 (20.7)
1a	88.9 (9)	98.7 (6.9)	101.7 (7.6)	98.6 (12.7)	89.5 (7.1)	108.1 (14.0)	102.7 (6.0)	112.0 (4.7)
2a	95.3 (4.1)	97.0 (6.2)	108.2 (16.9)	88.3 (17.4)	86.2 (5.0)	101.7 (15.7)	87.9 (7.1)	101.8 (11.1)
2b	28.0 (8.4) ****	19.7 (6.8) ****	36.8 (14.1) ****	44.4 (14.9) ****	30.5 (18.2) ****	45.0 (8.6) ****	73.7 (8.8)	95.3 (7.7)
2c	68.8 (7.3) ****	63.9 (16.2) ****	47.8 (6.4) ****	66.6 (9.5) ****	72.1 (12.7) ***	76.3 (13.7)	92.6 (13.6)	100.5 (18.8)
2d	103.6 (10.7)	111.1 (9.7)	105.1 (6.0)	104.0 (2.6)	106.9 (9.9)	92.5 (22.2)	99.7 (22.7)	111.8 (9.0)
2e	54.0 (10.5) ****	50.0 (14.2) ****	39.8 (1.8) ****	55.1 (11.5) ****	56.9 (7.6) ****	69.3 (5.3) **	64.1 (6.0) **	74.2 (7.9)
3	80.5 (10.4)	48.7 (7.8) ****	50.9 (6.4) ****	50.8 (9.7) ****	62.0 (4.2) ****	54.5 (9.2) ****	52.9 (2.4) ****	73.6 (18.9)
4	88.5 (7.9)	88.9 (11.1)	81.8 (15.5)	86.9 (7.6)	89.7 (11.6)	91.5 (21.5)	88.9 (5.2)	93.3 (7.2)
5a	16.1 (4.5) ****	8.6 (2.3) ****	6.4 (1.9) ****	11.8 (1.6) ****	11.6 (4.3) ****	29.6 (13.3) ****	38.5 (7.9) ****	14.3 (2.0) ****
5b	73.5 (8.7) *****	19.4 (2.1) ****	66.4 (3.3) ****	77.6 (11.6) ***	73.7 (2.4) ***	57.8 (16.6) ****	77.4 (9.8) #	n.d.
6	86.4 (10.6)	88.3 (10.7)	82.5 (4.7)	81.9 (10.6)	86.0 (15.3)	75.0 (11.8)	74.4 (14.8)	75.8 (13.6)
Crude extract	26.7 (0.6) ***	16.2 (4.0) ***	24.7 (6.7) ***	37.9 (7.2) ***	12.3 (2.4) ***	51.8 (8.7) ***	59.4 (2.0) ***	39.1 (9.7) ***

The results are the mean (SD) of at least four independent experiments, each in duplicate. Significant differences (** $p < 0.01$; *** $p < 0.001$ and **** $p < 0.0001$) when compared with control cells were evaluated by one-way ANOVA, followed by the post-hoc Dunnett's test. # Indicates significant differences when **5b** is compared with **5a**, as evaluated by a *t*-test. Compounds/extract are marked in light gray when cell viability is equal to or greater than 50% and in dark gray when cell viability is lower than 50% relative to control. n.d.—not determined.

Table 6. IC$_{50}$ values (half-maximal inhibitory concentration) and respective 95% confidence intervals of compounds tested in eight human cancer cell lines, determined by the MTT assay.

Compounds (µM)	IC$_{50}$ (95% CI)							
	HepG2	HT29	HCT116	A549	A375	MCF7	U251	T98G
2b	56.3 (37.9–83.7)	34.8 (22.0–55.0)	60.8 (41.6–88.8)	84.1 (60.3–117.1)	82.8 (50.5–123.8)	94.8 (63.2–142.3)	n.d.	n.d.
2c	123.8 (86.4–177.5)	113.7 (78.9–163.8)	72.9 (56.0–94.9)	109.0 (66.2–179.5)	146.4 (102.3–209.4)	n.d.	n.d.	n.d.
2e	118.9 (88.9–159.0)	111.6 (85.9–145.0)	73.2 (59.3–90.4)	105.8 (72.8–153.7)	123.1 (91.9–164.9)	186.8 (155.1–224.8)	212.5 (156.2–289.2)	n.d.
3	n.d.	43.2 (28.3–65.7)	64.1 (42.9–95.7)	85.2 (61.9–117.3)	109.8 (76.4–157.6)	99.7 (70.8–143.0)	120.2 (81.2–177.7)	n.d.
5a	43.2 (30.7–60.8)	12.3 (10.1–15.5)	29.8 (19.1–46.6)	47.9 (33.2–69.0)	38.5 (29.8–49.7)	53.6 (39.4–72.9)	74.1 (54–0–101.6)	50.4 (32.0–79.4)
5b	153.4 (96.2–244.7)	57.2 (41.7–78.4)	124.8 (94.9–164.1)	215.7 (161.4–288.1)	148.5 (103.7–212.4)	128.8 (94.5–175.5)	128.6 (86.7–190.8)	–
Dox	0.12 (0.07–0.22)	0.63 (0.26–1.12)	0.29 (0.16–0.54)	0.24 (0.13–0.08)	0.05 (0.03–0.08)	0.36 (0.16–0.84)	1.11 (0.43–2.85)	15.4 (10.4–22.9)

n.d.: IC$_{50}$ not determined because no cytotoxicity was observed at 100 µM. IC$_{50}$ values (in µM) are the mean of at least four independent experiments, each in duplicate. Doxorubicin (Dox) was used.

The results are the mean (SD) of at least four independent experiments, each in duplicate. Significant differences (** $p < 0.01$; *** $p < 0.001$ and **** $p < 0.0001$) when compared with control cells were evaluated by one-way ANOVA, followed by the post-hoc Dunnett's test. [#] Indicates significant differences when **5b** is compared with **5a**, as evaluated by a *t*-test. Compounds/extract are marked in light gray when cell viability is equal to or greater than 50% and in dark gray when cell viability is lower than 50% relative to control. n.d.—not determined.

3. Experimental Section

3.1. General Experimental Procedures

Melting points were determined on a Bock monoscope and are uncorrected. Optical rotations were measured on an ADP410 Polarimeter (Bellingham + Stanley Ltd., Tunbridge Wells, Kent, UK). Infrared spectra were recorded in a KBr microplate in a FTIR spectrometer Nicolet iS10 from Thermo Scientific (Waltham, MA, USA) with Smart OMNI-Transmission accessory (Software 188 OMNIC 8.3). ^{1}H and ^{13}C NMR spectra were recorded at ambient temperature on a Bruker AMC instrument (Bruker Biosciences Corporation, Billerica, MA, USA) operating at 300 or 500 and 75 or 125 MHz, respectively. High resolution mass spectra were measured with a Waters Xevo QToF mass spectrometer (Waters Corporations, Milford, MA, USA) coupled to a Waters Aquity UPLC system. A Merck (Darmstadt, Germany) silica gel GF_{254} was used for preparative TLC, and a Merck Si gel 60 (0.2–0.5 mm) was used for column chromatography.

3.2. Fungal Material

The strain KUFA 0062 was isolated from the marine sponge *Epipolasis* sp., which was collected, by scuba diving at a depth of 15–20 m from the coral reef at Similan Island National Park (8°39′09″ N, 97°38′27″ E), Phang-Nga province, Southern Thailand, in April 2010. The sponge was washed with 0.06% sodium hypochlorite solution for 1 min, followed by sterilized seawater three times, and then dried on sterile filter paper under a sterile aseptic condition. The sponge was cut into small pieces (5 × 5 mm), and placed on Petri dish plates containing 15 mL malt extract agar (MEA) medium with 70% seawater and 300 mg/L of streptomycin sulfate and incubated at 28 °C for 7 days. The hyphal tips emerging from the sponge pieces were individually transferred onto a MEA slant and maintained as pure culture for further identification.

The fungus was identified as *Aspergillus candidus*, based on morphological characters as described by Varga et al. [27].This identification was supported by molecular techniques using ITS primers. DNA was extracted from young mycelia following a modified Murray and Thompson method [28]. Primer pairs ITS1 and ITS4 [29] were used for ITS gene amplification. PCR reactions were conducted on Thermal Cycler and the amplification process consisted of initial denaturation at 95 °C for 5 min, 34 cycles at 95 °C for 1 min (denaturation), at 55 °C for 1 min (annealing) and at 72 °C for 1.5 min (extension), followed by a final extension at 72 °C for 10 min. PCR products were examined by Agarose gel electrophoresis (1% agarose with 1 × TBE buffer) and visualized under UV light after staining with ethidium bromide. DNA sequencing analyses were sequenced using the dideoxyribonucleotide chain termination method [30] by Macrogen Inc. (Seoul, Korea). The DNA sequences were edited using FinchTV software and submitted into the BLAST program for alignment and compared with that of fungal species in the NCBI database (http://www.ncbi.nlm.nih.gov/). Its gene sequences were deposited in GenBank with accession numbers KX 431210. The pure cultures were deposited as KUFA 0062 at Kasetsart University Fungal Collection, Department of Plant Pathology, Faculty of Agriculture, Kasetsart University, Bangkok, Thailand.

3.3. Extraction and Isolation

The fungus was cultured for one week at 28 °C in five Petri dishes (i.d. 90 mm) containing 15 mL of potato dextrose agar per dish. In order to obtain the mycelial suspension, the mycelial plugs were transferred to two 500 mL Erlenmeyer flasks containing 250 mL of potato dextrose broth, and then incubated on a rotary shaker at 150 rpm at 28 °C for 7 days. Fifty 1000 mL Erlenmeyer flasks, each containing 200 g of cooked rice, were autoclaved at 121 °C for 15 min, and then inoculated with 25 mL of mycelial suspension of *A. candidus*, and incubated at 28 °C for 30 days, after which the moldy rice was macerated in ethyl acetate (25 L total) for 7 days, and then filtered with Whatman No. 1 filter paper. The ethyl acetate solutions were combined and concentrated under reduced pressure to yield 53 g of crude ethyl acetate extract, which was dissolved in 1000 mL of $CHCl_3$, and filtered with Whatman No. 1 filter paper. The chloroform solution was then washed with H_2O (3 × 500 mL) and dried with anhydrous Na_2SO_4, filtered and evaporated under reduced pressure to give 50 g of the crude chloroform extract, which was applied on a column of silica gel (420 g) and eluted with mixtures of petrol-$CHCl_3$ and $CHCl_3$-Me_2CO, wherein the 250 mL fractions were collected as follow: Frs 1–80 (petrol-$CHCl_3$, 1:1), 81–220 (petrol-$CHCl_3$, 3:7), 221–350 (petrol-$CHCl_3$, 1:9), 351–579 ($CHCl_3$), 580–680 ($CHCl_3$-Me_2CO, 9.5:0.5), 681–730 ($CHCl_3$-Me_2CO, 9:1), 731–822 ($CHCl_3$-Me_2CO, 7:3). Frs 8–19 were combined (43.0 mg) and recrystallized in MeOH to give 17.4 mg of a yellow solid, which was identified as chrysopharic acid (**1a**) [13]. Frs 79–90 were combined (879.2 mg) and precipitated in MeOH to give 86.1 mg of clionasterol [23]. Frs. 91–118 were combined (2.49 g) and precipitated in MeOH to give 1.0 mg of palmitic acid. Frs 462–517 were combined (253.9 mg) and recrystallized in MeOH to give 31.4 mg of an orange solid of emodin (**1b**) [14]. Frs 528–583 were combined (535.8 mg) and precipitated in MeOH to give 254.4 mg of kumbicin D (**4**) [15], and the mother liquor that was crystallized in MeOH to give 11.8 mg of β-ergosterol 5,8-endoperoxide [14]. Frs 586–588 were combined (1.16 g) and recrystallized in Me_2CO to give 345.0 mg of white crystal of asterriquinol D dimethyl ether (**2a**) [15]. Frs 589–591 were combined (780.0 mg) and precipitated in Me_2CO to give a white solid, which was crystallized in MeOH to give 52.7 mg of kumbicin B (**2c**) [15], and the mother liquor was dried and recrystallized in Me_2CO to give 21.6 mg of candidusin D (**2e**). Frs 610–618 were combined (870.6 mg) and crystallized in a mixture of $CHCl_3$ and MeOH to give 250.6 mg of petromurin C (**2b**) [16]. The mother liquor of frs 610–618 (620.0 mg) was combined with frs. 594–609 (2.46 g) and frs 619–652 (1.29 g), and applied on a column chromatography of silica gel (130 g), and eluted with mixtures of petrol-$CHCl_3$ and $CHCl_3$-Me_2CO, wherein 250 mL sfrs were collected as follow: Sfrs 1–35 (petrol-$CHCl_3$, 3:7), 36–60 (petrol-$CHCl_3$, 1:9), 61–100 ($CHCl_3$), 101–120 ($CHCl_3$-Me_2CO, 9:1), 121–127 ($CHCl_3$-Me_2CO, 7:3). Sfrs 81–84 were combined (55.8 mg) and precipitated in Me_2CO to give 16.3 mg of 2″-oxoasterriquinol D methyl ether (**3**) [17]. Sfrs 85–89 were combined (102.3 mg) and precipitated in $CHCl_3$ to give 30.0 mg of petromurin C (**2b**). Frs 680–770 were combined (5.81 g) and applied on a column chromatography of silica gel (130 g), and eluted with mixtures of petrol-$CHCl_3$, $CHCl_3$-Me_2CO, wherein 250 mL fractions were collected as follow: Sfrs 1–30 (petrol-$CHCl_3$, 3:7), 31–77 (petrol-$CHCl_3$, 1:9), 78–101 ($CHCl_3$), 102–145 ($CHCl_3$-Me_2CO, 9:1). Sfrs 88–104 were combined (373.3 mg) and recrystallized in a mixture of $CHCl_3$ and Me_2CO to give 131.9 mg of preussin (**5a**) [18,19]. Frs 709–722 were combined (814.7 mg) and precipitated in MeOH to give 8 mg of (3*S*, 6*S*)-3,6-dibenzylpiperazine-2,5-dione (**6**). Frs 757–764 were combined (1.20 g) and precipitated in Me_2CO to give 7 mg of kumbicin A (**2d**). Frs 789–791 were combined (1.01 g) and applied over a column chromatography of Sephadex LH-20 (20 g) and eluted with MeOH, wherein 30 sfrs of 10 mL were collected. Sfrs 24–26 were combined (48.7 mg) and precipitated in Me_2CO to give 7.1 mg of 4-acetylamino benzoic acid (**7**) [22]. Frs 792–806 were combined (559.4 mg) and applied over a column chromatography of Sephadex LH-20 (20 g), and eluted with MeOH, wherein 40 sfrs of 10 mL were collected. Sfrs 8–14 were combined (350.4 mg) and precipitated in Me_2CO to 25 mg of preussin C (**5b**). Frs 807–822 were combined (416.1 mg) and precipitated in a mixed $CHCl_3$ and Me_2CO to give an additional 22.1 mg of preussin C (**5b**).

3.3.1. Candidusin D (**2e**)

White solid, m.p. 299–300 °C (CHCl$_3$/MeOH); IR (KBr) ν_{max} 3346, 2990, 2935, 1625, 1579, 1484, 1459, 1390, 1291 cm^{-1}; For ^{1}H and ^{13}C spectroscopic data (DMSO, 300.13 and 75.4 MHz), see Table 1; (+)-HRESIMS *m/z* 489.2030 (M + H)$^+$ (calculated for C$_{28}$H$_{29}$N$_2$O$_6$, 489.2026).

3.3.2. Preussin C (**5b**)

White solid, m.p. 219–220 °C (CHCl$_3$/MeOH); $[\alpha]_D^{23}$ + 28.6 (*c* 0.04, MeOH); IR (KBr) ν_{max} 3443, 2958, 2921, 2851, 2803, 2360, 2341,1651, 1634, 1581, 1470, 1411, 1261, 1095, 1031, 803 cm^{-1}; For ^{1}H and ^{13}C spectroscopic data, see Table 2; (+)-HRESIMS *m/z* 304.2647 (M + H)$^+$ (calculated for C$_{20}$H$_{34}$NO, 304.2640).

3.4. Electronic Circular Dichroism

Experimental ECD spectra were obtained with a Jasco J-815 CD spectropolarimeter in a 0.1 mm cuvette. The model construction, dihedral driver search, and MM2 minimizations were done in Chem3D Ultra (Perkin-Elmer Inc., Waltham, MA, USA). All other conformational energy minimization and spectral calculations (TDDFT method) were performed with Gaussian 16W (Gaussian Inc., Wallingford, CT, USA) at the PM6 or APFD/6-311+G(2d,p) level [31] with the IEFPCM methanol solvation model. The simulated spectral lines were obtained by summation of Gaussian curves, as recommended in [32]. ECD spectra were added using Boltzmann weights derived from its minimal APFD energies [33].

3.5. X-ray Crystal Structure of 3 and 5a

Single crystals were mounted on cryoloops using paratone. X-ray diffraction data were collected at room temperature with a Gemini PX Ultra equipped with CuK$_\alpha$ radiation (λ = 1.54184 Å). The structures were solved by direct methods using SHELXS-97 and refined with SHELXL-97 [34]. Non-hydrogen atoms were refined anisotropically. Hydrogen atoms were either placed at their idealized positions using appropriate HFIX instructions in SHELXL and included in subsequent refinement cycles or were directly found from difference Fourier maps and were refined freely with isotropic displacement parameters.

Full details of the data collection and refinement and tables of atomic coordinates, bond lengths and angles, and torsion angles have been deposited with the Cambridge Crystallographic Data Centre.

2″-Oxoasterriquinol D methyl ether (**3**). Crystal was triclinic, space group P-1, cell volume 1212.4(3) Å^3 and unit cell dimensions *a* = 9.364(2) Å, *b* = 11.3095(10) Å and *c* = 12.5789(12) Å and angles α = 105.760(8)°, β = 102.321(14)° and γ = 100.259(13)° (uncertainties in parentheses). There are two molecules per unit cell with calculated density of 1.212 g/cm^{-3}. Crystal has an inversion center (space group P-1) and thus the two molecules in the unit cell are enantiomers. The refinement converged to R (all data) = 15.23% and wR2 (all data) = 41.23%. (CCDC 1823071).

Preussin (**5a**). Crystal was triclinic, space group P1, cell volume 547.58(10) Å^3 and unit cell dimensions *a* = 5.8517(5) Å, *b* = 7.0327(6) Å and *c* = 14.2627(19) Å and angles α = 79.947(10)°, β = 79.515(9)° and γ = 73.094(7)° (uncertainties in parentheses). Calculated density of 1.073 g/cm^{-3}. The absolute structure was established with confidence (Flack parameter = 0.018(15)). The refinement converged to R (all data) = 3.42% and wR2 (all data) = 8.07%. (CCDC 1823438).

3.6. Antibacterial Activity Bioassays

3.6.1. Bacterial Strains and Growth Conditions

Four reference and three multidrug-resistant bacterial strains were used in this study. The Gram-positive bacteria comprised *Staphylococcus aureus* ATCC 29213, *Enterococcus faecalis* ATCC 29212, MRSA *S. aureus* 66/1 isolated from public buses [35], and VRE *E. faecalis* B3/101 isolated from river water [36]. The Gram-negative bacteria used were *Escherichia coli* ATCC 25922, *Pseudomonas aeruginosa* ATCC 27853, a colistin-resistant *E. coli* 1418/1 strain isolated from rabbit feces, and a clinical isolate ESBL *E. coli* SA/2. Frozen stocks of all strains were grown in Mueller-Hinton agar (MH-BioKar diagnostics, Allone, France) at 37 °C. All bacterial strains were sub-cultured in MH agar and incubated overnight at 37 °C before each assay.

3.6.2. Antimicrobial Susceptibility Testing

The minimum inhibitory concentration (MIC) was used for determining the antibacterial activity of each compound, in accordance with the recommendations of the Clinical and Laboratory Standards Institute (CLSI) [37]. With the exception of **6**, 10 mg/mL stock solutions of each compound were prepared in dimethylsulfoxide (DMSO—Applichem GmbH, Darmstadt, Germany). For **6**, which was less soluble in DMSO than the other compounds, a stock solution of 1 mg/mL was prepared. Two-fold serial dilutions of the compounds were prepared in Mueller-Hinton broth 2 (MHB2—Sigma-Aldrich, St. Louis, MO, USA) within the concentration range of 0.062–64 µg/mL, except for **6** for which the highest concentration tested was 32 µg/mL. The highest concentration tested was chosen in order to maintain DMSO in-test concentration below 1% as recommended by the CLSI [37]. At this concentration DMSO did not affected bacterial growth. Cefotaxime (CTX) ranging between 0.031 and 16 µg/mL was used as a control. A purity check and colony counts of the inoculum suspensions were also evaluated in order to ensure that the final inoculum density closely approximates the intended number (5×10^5 CFU/mL). The MIC was determined as the lowest concentration at which no visible growth was observed. The minimum bactericidal concentration (MBC) was assessed by spreading 10 µL of culture collected from wells showing no visible growth on MH agar plates. The MBC was determined as the lowest concentration at which no colonies grew after 16–18 h incubation at 37 °C. These assays were performed in duplicate.

3.6.3. Biofilm Formation Inhibition Assay

The effect of all compounds on biofilm formation was assessed using crystal violet staining as previously described [38]. Briefly, the highest concentration tested in the MIC assay was added to bacterial suspensions of 1×10^6 CFU/mL prepared in Tryptic Soy broth (TSB-BioKar diagnostics, Allonne, France). When it was possible to determine a MIC, four concentrations were tested: $2 \times$ MIC, MIC, $\frac{1}{2}$ MIC, and $\frac{1}{4}$ MIC. A control without any compound as well as a negative control (TSB alone) were included. The stabilized biofilm mass was quantified after 24 h incubation at 37 °C. The absorbance was measured at 595 nm on an iMark™ microplate spectrophotometer (Bio-Rad Laboratories, Hercules, CA, USA). The background absorbance (TSB without inoculum) was subtracted from the absorbance of each sample and data are shown as a percentage of control. Three independent experiments were performed in triplicate for each experimental condition.

3.6.4. Antibiotic Synergy Testing

In order to evaluate the combined effect of the compounds and clinically relevant antimicrobial drugs, a screening was conducted using the disk diffusion method, as previously described [38,39]. A set of antibiotic disks (Oxoid, Basingstoke, UK) to which the isolates were resistant was selected: cefotaxime (CTX, 30 µg) for *E. coli* SA/2, oxacillin (OX, 1 µg) for *S. aureus* 66/1, and vancomycin (VA, 30 µg) for *E. faecalis* B3/101. Antibiotic disks alone (controls) and antibiotic disks impregnated with 15 µg of each compound were placed on MH agar plates seeded with the respective bacteria. Sterile 6 mm blank paper disks (Oxoid, Basingstoke, UK) impregnated with 15 µg of each compound alone were also tested. A blank disk with DMSO was used as a negative control. MH inoculated plates were incubated for 18–20 h at 37 °C. Potential synergism was recorded when the halo of an antibiotic disk impregnated with a compound was greater than the halo of the antibiotic or compound-impregnated blank disk alone.

The potential synergy between the most promising compounds and clinically relevant antibiotics (oxacillin and vancomycin—Sigma-Aldrich, St. Louis, MO, USA) was also evaluated by the checkerboard method as previously described [38]. For this purpose multidrug-resistant strains were selected based on their resistance towards those antibiotics. The fractional concentration (FIC) was calculated as follows: FIC of drug A (FIC A) = MIC of drug A in combination/MIC of drug A alone, and FIC of drug B (FIC B) = MIC of drug B in combination/MIC of drug B alone. The FIC index was calculated as the sum of each FIC and interpreted as follows: $\Sigma FIC \leq 0.5$, synergy; $0.5 < \Sigma FIC \leq 4$, no interaction; $4 < \Sigma FIC$, antagonism [40]. When it was not possible to determine a MIC value for the test compound, the potential synergy between the compounds and clinically relevant antibiotics was evaluated by determining the antibiotic MIC in the presence of each compound. Briefly, the MIC of CTX, OX, VA, and colistin (Sigma-Aldrich, St. Louis, MO, USA) for the respective multidrug-resistant strains was determined in the presence of the highest concentration of each compound tested, unless otherwise stated, that did not affect bacterial growth when the compound was used alone. The antibiotic tested was serially diluted whereas the concentration of each compound was kept fixed. In the case of **6** the concentration used was 32 µg/mL. For all other compounds the concentration used was 64 µg/mL. Antibiotic MICs where determined as described above.

3.7. In Vitro Anticancer Activity Assays

3.7.1. Cell Lines

Eight cancer cell lines were used to assess the anticancer activity of the extract and isolated compounds of the marine sponge-associated fungus *Aspergillus candidus* KUFA 0062. HT29, HCT116 cells were kindly provided by Prof. Carmen Jerónimo, from CI-IPO, Porto and HepG2 cells were provided by Prof. Rosário Martins, from ESTSP and CIIMAR, Porto. A375, A549, MCF7, U-251 and T98G cell lines were obtained from the European Collection of Cell Cultures. Cells were maintained as monolayer cultures in DMEM supplemented with 10% FBS, 1% antibiotic solution (100 U/mL penicillin and 100 µg/mL streptomycin), 10 mM N-[2-hydroxyethyl] piperazine-N'-[2-ethane-sulfonic acid] and 0.1 mM sodium pyruvate in a humidified incubator with 5% CO_2 at 37 °C. Cells were trypsinized near the confluence. Stock solutions of the isolated compounds, extract and Dox were prepared in dimethyl sulphoxide (DMSO) and aliquots and kept at −20 °C. For experiments, the final concentration of DMSO in the medium was < 0.1% (v/v) and the controls received DMSO only.

3.7.2. MTT Reduction Assay

To assess the effects of the isolated compounds and the extract on cell viability, cells were plated in 96-multiwell culture plates at a density of 0.1×10^6 cells/mL. After cell adhesion, cells were incubated for 48 h with fresh medium containing the isolated compounds, at 100 µM, and extract, at 200 µg/mL, and cell viability was measured by MTT reduction assay as described previously [41]. Briefly, after 48 h of treatment MTT was added at a final concentration of 0.5 mg/mL and incubated for 2 h at 37 °C. The formazan crystals were dissolved in a DMSO: EtOH solution (1:1) (v/v) and the absorbance was measured at 570 nm in a microplate reader (Multiskan Ex, Labsystems, Milford, MA, USA).

The determination of IC_{50} (half-maximal inhibitory concentration) was performed only for the isolated compounds that significantly decrease cell viability at 100 µM. For this, cancer cell lines were incubated with different concentrations of isolated compounds (0–200 µM) or Dox (0.001–10 µM), as a positive control, and the MTT assay was performed as described above. The IC_{50} values were determined using GraphPad Prism v6.0 software (GraphPad Software, La Jolla, CA, USA). This software was also used to conduct the statistical analysis, using a one-way ANOVA, followed by the post-hoc Dunnett's test. The normality and homogeneity of variance were confirmed by the Shapiro-Wilk test and Levine test, respectively. The significance level was set at the conventional 95%.

4. Conclusions

Most of the investigations of the secondary metabolites of the fungus *Aspergillus candidus* were carried out with its terrestrial strains. Except for a chlorine containing flavone antifungal antibiotic chlorflavonin, the secondary metabolites produced by this fungal species were *p*-terphenyl and *bis*-indolyl benzenoids derivatives, some of which immunosuppressive activity, cytotoxicity against sea urchin embryos and various tumor cell lines. Interestingly, to the best of our knowledge, only two marine-derived *A. candidus* were investigated chemically. While *A. candidus*, isolated from the marine sediment furnished *p*-terphenyl derivatives, the strain isolated from the gut of sea urchin produced a different group of metabolites, i.e., spiculisporic derivatives. This fact has prompted us to investigate the secondary metabolites from a marine-sponge associated strain of *A. candidus* to verify if the marine host can influence the type of secondary metabolites produced by this fungus. Our results were very interesting since we isolated, besides the commonly found fungal metabolites such as palmitic acid, ergosterol-5,8-endoperoxide, the anthraquinones chrysophanic acid and emodin, a diketopiperazine derivative, and *bis*-indolyl derivatives that are the chemical signature of this fungus, one previously reported hydroxypyrrolidine alkaloid preussin (**5a**) and a new congener, preussin C (**5b**). Although preussin (**5a**) was isolated from *A. ochraceus* and other fungi (*Preussia* sp. and *Simplicillium lanosoniveum*), this is the first report of the isolation of hydroxypyrrolidine alkaloids from *A. candidus*. From the biological perspective, it is interesting to note that although most of the tested secondary metabolites exhibited a cytotoxic effect against eight human cancer cell lines, their activity varied depending on the cell lines. The most notable aspect was that the presence of the *N*-methyl group on the pyrrolidine ring in preussin (**5a**) was always crucial in both antibacterial and cytotoxic activities. This was demonstrated by the effectiveness of preussin (**5a**) to decrease cell viability when compared to the *N*-demethyl analogue, preussin C (**5b**). For the antibacterial activity assay, it is important to point out that, among the compounds tested, only preussin (**5a**) stood out since it displayed a relevant inhibitory effect against Gram-positive bacteria as well as methicillin-resistant *S. aureus* (MRSA) and vancomycin-resistant enterococci (VRE) strains. Moreover, preussin (**5a**) was not only able to interfere with biofilm formation in *S. aureus* ATCC 29213 and *E. faecalis* ATCC 29212 at concentrations equal to or above the MIC but also showed synergistic effect with vancomycin and oxacillin against VRE and MRSA isolates and a strong synergistic effect with colistin when tested against *E. coli* 1410/1. Therefore, our study reveals that preussin (**5a**) can represent a new model for the development of a new class of antibiotic and anticancer agents.

Supplementary Materials: The following are available online at http://www.mdpi.com/1660-3397/16/4/119/s1, Figure S1: Structures of palmitic acid, clionasterol and ergosterol-5,8-endoperoxide, isolated from the marine sponge-associated fungus *Aspergillus candidus* KUFA0062, Figures S2–S35: 1D and 2D NMR spectra of isolated compounds, Table S1: ¹H NMR data of **2a–d** (DMSO, 300.13 MHz), Table S2: ¹³C NMR data of **2a–d** (DMSO, 75.3 MHz), Table S3: Comparison of ¹H and ¹³C NMR data of **3** (DMSO, 300.13 and 75.4 MHz) and 2″-oxoasterriquinol D methyl ether (CDCl₃, 300 MHz), Table S4: ¹H and ¹³C NMR data of **4** (kumbicin D) (300.13 MHz and 75.4 MHz).

Acknowledgments: This work was partially supported through national funds provided by the FCT/MCTES-Foundation for Science and Technology from the Minister of Science, Technology and Higher Education (PIDDAC) and European Regional Development Fund (ERDF) through the COMPETE—Programa Operacional Factores de Competitividade (POFC) programme, under the project PTDC/MAR-BIO/4694/2014 (reference POCI-01-0145-FEDER-016790; Project 3599-Promover a Produção Científica e Desenvolvimento Tecnológico e a Constituição de Redes Temáticas (3599-PPCDT)) in the framework of the program PT2020 as well as by the project INNOVMAR-Innovation and Sustainability in the Management and Exploitation of Marine Resources (reference NORTE-01-0145-FEDER-000035, within Research Line NOVELMAR), supported by North Portugal Regional Operational Program (NORTE 2020), under the PORTUGAL 2020 Partnership Agreement, through the European Regional Development Fund (ERDF). We thank Júlia Bessa and Sara Cravo for technical support.

Author Contributions: Anake Kijjoa, Madalena M. M. Pinto and José A. Pereira conceived of and designed the experiment and elaborated the manuscript; Suradet Buttachon performed the isolation and purification of the compounds; Tida Dethoup collected, isolated, identified, and cultured the fungus; Luís Gales performed X-ray analysis; José A. Pereira performed calculations and measurement of the ECD spectra. Paulo M.

Costa and Ângela Inácio performed and interpreted the results of antibacterial assays; Alice A. Ramos and Eduardo Rocha performed and interpreted the results of in vitro anticancer assays; Nazim Sekeroglu assisted in the elaboration of the manuscript; Michael Lee provided HRMS; Artur M. S. Silva provided the NMR spectra.

Conflicts of Interest: The authors declare no conflict of interest.

References

1. Samson, R.A.; Visagie, C.M.; Houbraken, J.; Hong, S.B.; Hubka, V.; Klaassen, C.H.; Perrone, G.; Seifert, K.A.; Susca, A.; Tanney, J.B.; et al. Phylogeny, identification and nomenclature of the genus *Aspergillus*. *Stud. Mycol.* **2014**, *78*, 141–173. [CrossRef] [PubMed]

2. Park, J.W.; Choi, S.Y.; Hwang, H.J.; Kim, Y.B. Fungal mycoflora and mycotoxins in Korean polished rice destined for humans. *Int. J. Food Microbiol.* **2005**, *103*, 305–314. [CrossRef] [PubMed]

3. Ismail, M.A.; Taligoola, H.K.; Chebon, S.K. Mycobiota associated with rice grains marketed in Uganda. *J. Biol. Sci.* **2004**, *4*, 271–278.

4. Kozakiewicz, Z. *Aspergillus* species on stored products. *Mycol. Pap.* **1989**, *161*, 1–188.

5. Richards, M.; Bird, A.E.; Munden, J.E. Chloflavonin a new fungal antibiotic. *J. Antibiot.* **1969**, *22*, 388–389. [CrossRef]

6. Marchelli, R.; Vining, L.C. Terphenyllin, a novel *p*-terphenyl metabolite from *Aspergillus candidus*. *J. Antibiot.* **1975**, *23*, 328–331. [CrossRef]

7. Takahashi, C.; Yoshihira, K.; Natori, S.; Umeda, M. The structure of toxic metabolites of *Aspergillus candidus* I. The compounds A and E, cytotoxic *p*-terphenyls. *Chem. Pharm. Bull.* **1976**, *24*, 613–620. [CrossRef] [PubMed]

8. Kurobane, I.; Vining, L.C.; McInnes, A.G.; Smith, D.G. 3-Hydroxyterphenyllin, a new metabolite of *Aspergillus candidus*. *J. Antibiot.* **1979**, *32*, 559–564. [CrossRef] [PubMed]

9. Kobayashi, A.; Takemura, A.; Nagano, H.; Kawazu, K. Candidusins A and B: New *p*-terphenyls with cytotoxic effects on sea urchin embryos. *Agric. Biol. Chem.* **1982**, *6*, 585–589. [CrossRef]

10. Kamigauchi, T.; Sakazaki, R.; Nagashima, K.; Kawamura, Y.; Yasuda, Y.; Matsushima, K.; Tani, H.; Takahashi, Y.; Suzuki, R.; Koizumi, K.; et al. Terprenins novel immunosuppressants produced by *Aspergillus candidus*. *J. Antibiot.* **1998**, *51*, 445–450. [CrossRef] [PubMed]

11. Wei, H.; Inada, H.; Hayashi, A.; Higashimoto, K.; Pruksakorn, P.; Kamada, S.; Arai, A.; Ishida, S.; Kobayashi, M. Prenylterphenyllin and its dehydroxyl analogs, new cytotoxic substances from marine-derived fungus *Aspergillus candidus* IF10. *J. Antibiot.* **2007**, *60*, 586–590. [CrossRef] [PubMed]

12. Wang, R.; Guo, Z.K.; Li, X.M.; Chen, F.X.; Zhan, X.F.; Shen, M.H. Spiculisporic acid analogues of the marine-derived fungus *Aspergillus candidus* strain HDf2, and their antibacterial activity. *Antonie Leeuwenhoek* **2015**, *108*, 215–219. [CrossRef] [PubMed]

13. Semple, S.J.; Pyke, S.M.; Reynolds, G.D.; Flower, R.L. In vitro antiviral activity of the anthraquinone chrysophanic acid against poliovirus. *Antivir. Res.* **2001**, *49*, 169–178. [CrossRef]

14. Noinart, J.; Buttachon, S.; Dethoup, T.; Gales, L.; Pereira, J.A.; Urbatzka, R.; Freitas, S.; Lee, M.; Silva, A.M.S.; Pinto, M.M.M.; et al. A new ergosterol analog, a new bis-anthraquinone and anti-obesity activity of anthraquinones from the marine sponge-associated fungus *Talaromyces stipitatus* KUFA 0207. *Mar. Drugs* **2017**, *15*, 139. [CrossRef] [PubMed]

15. Lacey, H.J.; Vuong, D.; Pitt, J.; Lacey, E.; Piggott, A.M. Kumbicins A–D: Bis-indolyl benzenoids and benzoquinone from an Australian soil fungus, *Aspergilus kumbius*. *Aust. J. Chem.* **2016**, *69*, 152–160. [CrossRef]

16. Ooike, M.; Nozawa, K.; Udagawa, S.; Kawai, K. Bisindolylbenzenoids from ascostromata of *Petromyces muricatus*. *Can. J. Chem.* **1997**, *75*, 625–628. [CrossRef]

17. Whyte, A.C.; Joshi, B.K.; Gloer, J.B.; Wicklow, D.T.; Dowd, P.F. New cyclic peptide and bisindolyl benzenoid metabolites from the sclerotia of *Aspergillus sclerotiorum*. *J. Nat. Prod.* **2000**, *63*, 1006–1009. [CrossRef] [PubMed]

18. Schwartz, R.E.; Liesch, J.; Hensens, O.; Zitano, L.; Honeycutt, S.; Garrity, G.; Fronmtling, R.A.; Onishi, J.; Monaghan, R. L-657,398, a novel antifungal agent: Fermentation, isolation, structural elucidation and biological properties. *J. Antibiot.* **1988**, *41*, 1774–1779. [CrossRef] [PubMed]

19. Johnson, J.H.; Phillipson, D.W.; Kahle, A.D. The relative and absolute stereochemistry of the antifungal preussin. *J. Antibiot.* **1989**, *42*, 1184–1185. [CrossRef] [PubMed]

20. Fukuda, T.; Sudoh, Y.; Tsuchirya, Y.; Okuda, T.; Igarashi, Y. Isolation and biosynthesis of preussin B, a pyrrolidine alkaloid from *Simplicillium lanosoniveum*. *J. Nat. Prod.* **2014**, *77*, 813–817. [CrossRef] [PubMed]

21. Wang, J.M.; Ding, G.Z.; Fang, L.; Dai, J.G.; Yu, S.S.; Wang, Y.H.; Chen, X.G.; Ma, S.G.; Qu, J.; Xu, S.; et al. Thiodiketopiperazines produced by the endophytic fungus *Epicoccum nigrum*. *J. Nat. Prod.* **2010**, *73*, 1240–1249. [CrossRef] [PubMed]

22. Parish, C.A.; Huber, J.; Baxter, J.; González, A.; Collado, J.; Platas, G.; Diez, M.T.; Vicente, F.; Dorso, K.; Abruzzo, G.; et al. A new ene-triyne antibiotic from the fungus *Baeospora myosura*. *J. Nat. Prod.* **2004**, *67*, 1900–1902. [CrossRef] [PubMed]

23. Dzeha, T.; Jaspars, M.; Tabudravu, J. Clionasterol, a triterpenoid from the Kenyan marine green macroalga *Halimeda macroloba*. *Western Indian Ocean J. Mar. Sci.* **2003**, *2*, 157–161.

24. Cantu, M.D.; Hillebrand, S.; Carrilho, E. Determination of the dissociation constants (pK_a) of secondary and tertiary amines in organic media by capillary electrophoresis and their role in the electrophoretic mobility order inversion. *J. Chromatogr. A* **2005**, *1068*, 99–105. [CrossRef] [PubMed]

25. EUCAST. Recommendations for MIC Determination of Colistin (Polymyxin E) as Recommended by the Joint CLSI-EUCAST Polymyxin Breakpoints Working Group. Available online: http://www.bioconnections.co.uk/files/merlin/Recommendations_for_MIC_determination_of_colistin_March_2016.pdf (accessed on 1 March 2018).

26. Ravizza, R.; Gariboldi, M.B.; Passarelli, L.; Monti, E. Role of the p53/p21 system in the response of human colon carcinoma cells to doxorubicin. *BMC Cancer* **2004**, *4*, 92. [CrossRef] [PubMed]

27. Varga, J.; Frisvad, J.C.; Samson, R.A. Polyphasic taxonomy of *Aspergillus* section *Candidi* based on molecular, morphological and physiological data. *Stud. Mycol.* **2007**, *59*, 75–88. [CrossRef] [PubMed]

28. Murray, M.G.; Thompson, W.F. Rapid isolation of high molecular weight plant DNA. *Nucleic Acids Res.* **1980**, *8*, 4321–4325. [CrossRef] [PubMed]

29. White, T.J.; Bruns, T.; Lee, S.; Taylor, J. Amplification and direct sequencing of fungal ribosomal RNA genes for phylogenetics. In *PCR Protocols: A Guide to Methods and Applications*; Innis, M.A., Gelfand, D.H., Sninsky, J.J., White, T.J., Eds.; Academic Press: New York, NY, USA, 1990; pp. 315–322.

30. Sanger, F.; Nicklen, S.; Coulson, A.R. DNA sequencing with chain-terminating inhibitors. *Proc. Natl. Acad. Sci. USA* **1977**, *72*, 5463–5467. [CrossRef]

31. Austin, A.; Petersson, G.A.; Frisch, M.J.; Dobek, F.J.; Scalmani, G.; Throssel, K. A density functional with spherical atom dispersion terms. *J. Chem. Theory Comput.* **2012**, *8*, 4989–5007. [CrossRef] [PubMed]

32. Stephens, P.J.; Harada, N. ECD Cotton effect approximated by the Gaussian curve and other methods. *Chirality* **2010**, *22*, 229–233. [CrossRef] [PubMed]

33. Mori, T.; Inoue, Y.; Grimme, S. Time dependent density functional theory calculations for electronic circular dichroism spectra and optical rotations of conformationally flexible chiral donor-acceptor dyad. *J. Org. Chem.* **2006**, *71*, 9797–9806. [CrossRef] [PubMed]

34. Sheldrick, G.M. A short story of SHELX. *Acta Cryst.* **2008**, *A64*, 112–122. [CrossRef] [PubMed]

35. Simões, R.R.; Aires-de-Sousa, M.; Conceicao, T.; Antunes, F.; da Costa, P.M.; de Lencastre, H. High prevalence of EMRSA-15 in Portuguese public buses: A worrisome finding. *PLoS ONE* **2011**, *6*, e17630. [CrossRef] [PubMed]

36. Bessa, L.J.; Barbosa-Vasconcelos, A.; Mendes, A.; Vaz-Pires, P.; Martins da Costa, P. High prevalence of multidrug-resistant *Escherichia coli* and *Enterococcus* spp. in river water, upstream and downstream of a wastewater treatment plant. *J. Water Health* **2014**, *12*, 426–435. [CrossRef] [PubMed]

37. Clinical and Laboratory Standards Institute. Methods for dilution antimicrobial susceptibility tests for bacteria that grow aerobically. In *Approved Standard*, 10th ed.; CLSI: Wayne, PA, USA, 2015; Document M07-A10.

38. May Zin, W.W.; Buttachon, S.; Dethoup, T.; Pereira, J.A.; Gales, L.; Inácio, A.; Costa, P.M.; Lee, M.; Sekeroglu, N.; Silva, A.M.S.; et al. Antibacterial and antibiofilm activities of the metabolites isolated from the culture of the mangrove-derived endophytic fungus *Eurotium chevalieri* KUFA 0006. *Phytochemistry* **2017**, *141*, 86–97. [CrossRef] [PubMed]

39. Bessa, L.J.; Palmeira, A.; Gomes, A.S.; Vasconcelos, V.; Sousa, E.; Pinto, M.; Martins da Costa, P. Synergistic effects between thioxanthones and oxacillin against methicillin-resistant *Staphylococcus aureus*. *Microb. Drug Resist.* **2015**, *21*, 404–415. [CrossRef] [PubMed]

40. Odds, F.C. Synergy, antagonism, and what the chequer board puts between them. *J. Antimicrob. Chemother.* **2003**, *52*. [CrossRef] [PubMed]

41. Ramos, A.A.; Prata-Sena, M.; Castro-Carvalho, B.; Dethoup, T.; Buttachon, B.; Kijjoa, A.; Rocha, E. Testing the potential of four marine-derived fungi extracts as anti-proliferative and cell death-inducing agents in seven cancer cell lines. *Asian Pac. J. Trop. Med.* **2015**, *8*, 798–806. [CrossRef] [PubMed]

marine drugs

MDPI

Article

Cytotoxic and Antibacterial Compounds from the Coral-Derived Fungus *Aspergillus tritici* SP2-8-1

Weiyi Wang [1,2,3,†], Yanyan Liao [1,2,4,†], Chao Tang [1,2,4], Xiaomei Huang [1,2,4], Zhuhua Luo [3], Jianming Chen [3,5,*] and Peng Cai [1,2,4,*]

[1] Key Laboratory of Urban Environment and Health, Institute of Urban Environment, Chinese Academy of Sciences, Xiamen 361021, China; wywang@iue.ac.cn (W.W.); yyliao@iue.ac.cn (Y.L.); ctang@iue.ac.cn (C.T.); xmhuang@iue.ac.cn (X.H.)

[2] University of Chinese Academy of Sciences, Beijing 100049, China

[3] State Key Laboratory Breeding Base of Marine Genetic Resources, Key Laboratory of Marine Genetic Resources, Fujian Key Laboratory of Marine Genetic Resources, Fujian Collaborative Innovation Centre for Exploitation and Utilization of Marine Biological Resources, Third Institute of Oceanography, State Oceanic Administration, Xiamen 361005, China; luozh_fj@163.com

[4] Xiamen Key Laboratory of Physical Environment, Xiamen 361021, China

[5] Institute of Oceanography, Minjiang University, Fuzhou 350108, China

* Correspondence: chenjianming@tio.org.cn (J.C.); pcai@iue.ac.cn (P.C.); Tel.: +86-592-219-5518 (J.C.); +86-592-619-0568 (P.C.)

† These authors contributed equally to this work.

Received: 1 September 2017; Accepted: 3 November 2017; Published: 7 November 2017

Abstract: Three novel compounds, 4-methyl-candidusin A (**1**), aspetritone A (**2**) and aspetritone B (**3**), were obtained from the culture of a coral-derived fungus *Aspergillus tritici* SP2-8-1, together with fifteen known compounds (**4–18**). Their structures, including absolute configurations, were assigned based on NMR, MS, and time-dependent density functional theory (TD-DFT) ECD calculations. Compounds **2** and **5** exhibited better activities against methicillin-resistant strains of *S. aureus* (MRSA) ATCC 43300 and MRSA CGMCC 1.12409 than the positive control chloramphenicol. Compound **5** displayed stronger anti-MRSA and lower cytotoxic activities than **2**, and showed stronger antibacterial activities against strains of *Vibrio vulnificus*, *Vibrio rotiferianus*, and *Vibrio campbellii* than the other compounds. Compounds **2** and **10** exhibited significantly stronger cytotoxic activities against human cancer cell lines HeLa, A549, and Hep G2 than the other compounds. Preliminary structure–activity relationship studies indicated that prenylation of terphenyllin or candidusin and the tetrahydrobenzene moiety in anthraquinone derivatives may influence their bioactivity.

Keywords: *Aspergillus*; candidusin; aspetritone; cytotoxic; antibacterial

---

1. Introduction

To date, approximately 70,000 species of fungi have been characterized [1]. Among them, about 1500 species of marine-derived fungi were mentioned, mainly from coastal ecosystems [1]. In recent years, the fungal sources of novel metabolites have broadened from saprophytic terrestrial strains to marine habitats and living plants with their endophytes [2]. Specifically, metabolites isolated from species of the genus *Aspergillus* have continually attracted the interest of pharmacologists due to their broad array of biological activities and their structural diversity. *A. tritici*, *A. campestris*, *A. taichungensis*, and *A. candidus*, which are members of the *Aspergillus* section *Candidi*, are known to be the prolific producers of bioactive secondary metabolites, including terphenyllin, candidusins, and anthraquinones [3].

As part of our ongoing efforts to discover bioactive compounds from coral-derived microorganisms, an *Aspergillus tritici* strain, SP2-8-1, isolated from the coral of *Galaxea fascicularis*, collected at Port Dickson, Malaysia, attracted our attention. Studies on the bioactive constituents of its extract led to the isolation of three novel compounds, 4-methyl-candidusin A (**1**), aspetritone A (**2**) and aspetritone B (**3**), together with fifteen known analogues, including prenylcandidusin derivatives (**4–5**), candidusin derivatives (**6–7**), terphenyllin derivatives (**8–14**), and anthraquinone derivatives (**15–18**) (Figure 1).

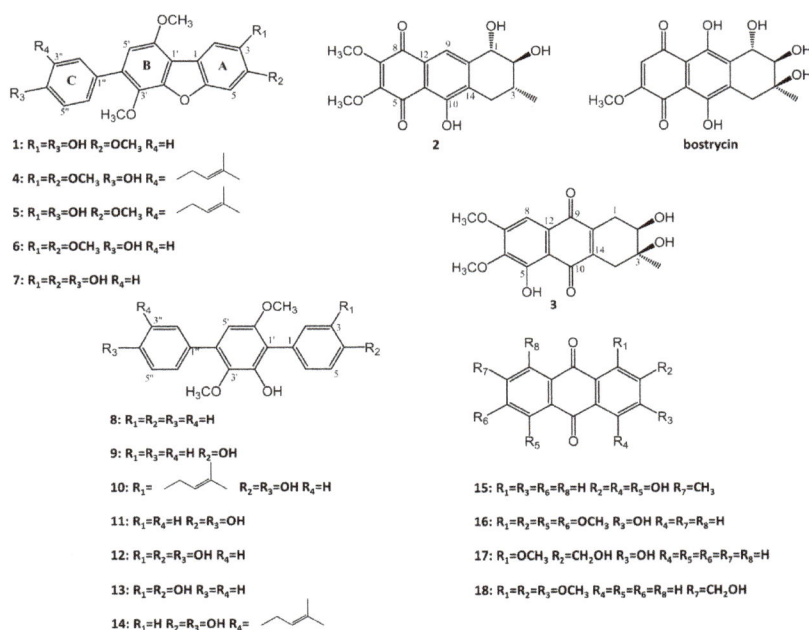

Figure 1. Structure of compounds **1–18**.

2. Results and Discussion

4-methyl-candidusin A (**1**) was obtained as a colorless amorphous solid. Its molecular formula was established as $C_{21}H_{18}O_6$ by high-resolution electrospray ionization mass spectroscopy (HR-ESI-MS) (m/z 367.11757 [M + H]$^+$; calcd for $C_{21}H_{19}O_6$, 367.11816), implying 13 degrees of unsaturation. The ^{13}C/distortionless enhancement by polarization transfer (DEPT) spectrum showed resonances for three methoxyl, seven methine, and 11 quaternary carbons. The ^1H NMR spectrum displayed an AB system at δ_H [6.85 (d, J = 8.53, H-3″, 5″) and 7.42 (d, J = 8.53, H-2″, 6″)]; three aromatic singlets at δ_H [7.39 (s, H-2)], δ_H [7.38 (s, H-5)], and δ_H [6.72 (s, H-5′)]; three methoxyl singlets at δ_H [3.87 (s, OCH$_3$-4)], δ_H [3.77 (s, OCH$_3$-3′)], and δ_H [3.97 (s, OCH$_3$-6′)]; and two phenolic OH groups at δ_H [9.06 (brs, OH-3)] and δ_H [9.55 (brs, OH-4″)] (Table 1). In comparison with the previously reported candidusin A [4], the lack of a phenolic OH unit and the appearance of a methoxyl group in **1** were observed, confirmed by evidence of a 14 amu increase in the molecular weight of **1**. In addition, they shared the same substructures of rings B and C, with the main differences located on ring A. In combining the correlations of ^1H–^1H COSY and heteronuclear multiple bond correlation (HMBC) spectra (Figure 2) with the "no splitting" of H-2 and H-5, we assigned the structure of compound **1** as 4-methyl-candidusin A (**1**).

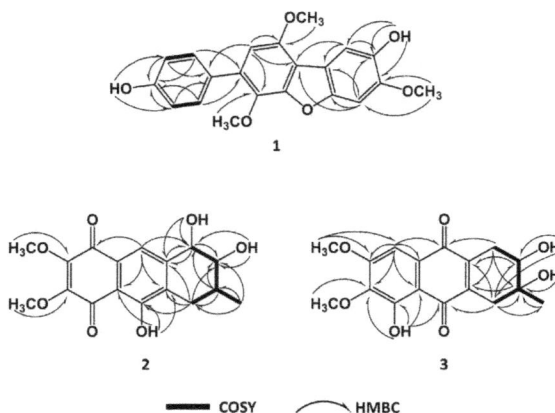

Figure 2. Key COSY and HMBC correlations of compounds **1–3**.

Table 1. ^{1}H NMR data (400 MHz) and ^{13}C NMR data (100 MHz) for compounds **1–3**.

Position	1		2		3	
	δ_H, mult. (*J* in Hz)	δ_C	δ_H, mult. (*J* in Hz)	δ_C	δ_H, mult. (*J* in Hz)	δ_C
1		114.9	4.26, d (7.03)	73.8	2.54, dd (19.84, 4.88), Heq 2.69, dd (19.84, 11.63), Hax	30.1
2	7.39, s	107.3	3.13, m	76.3	3.59, m	69.7
3		144.0	1.76, m	33.5		69.1
4		148.4	2.19, dd (18.45, 11.46), Hax 2.92, dd (18.45, 5.40), Heq	31.3	2.45, br d (13.30)	33.0
5	7.38, s	96.6		187.2		155.1
6		149.7		148.8		140.8
7				146.9		157.9
8				181.1	7.16, s	104.0
9			7.65, s	118.1		183.5
10				158.0		189.2
11				111.0		110.8
12				128.5		127.9
13				148.6		142.8
14				131.7		142.6
1'		114.2				
2'		149.0				
3'		136.4				
4'		131.4				
5'	6.72, s	106.0				
6'		150.0				
1''		129.0				
2'', 6''	7.42, d (8.53)	130.8				
3'', 5''	6.85, d (8.53)	115.5				
4''		157.2				
CH$_3$-3			1.08, d (6.53)	18.5	1.21, s	26.0
OCH$_3$-4	3.87, s	56.4				
OCH$_3$-6			3.99, s	61.4	3.79, s	60.8
OCH$_3$-7			3.95, s	61.7	3.92, s	56.8
OCH$_3$-3'	3.77, s	61.0				
OCH$_3$-6'	3.97, s	56.3				
OH-1			5.82, d (6.27)			
OH-2			5.06, d (5.02)		4.73, br s	
OH-3	9.06, brs				5.03, br s	
OH-5					12.09, s	
OH-10			12.18, s			
OH-4''	9.55, brs					

Aspetritone A (**2**) was obtained as a yellow amorphous solid. Its molecular formula was established as $C_{17}H_{18}O_7$ by HRESIMS (*m/z* 333.0966 [M − H]$^-$; calcd. for $C_{17}H_{17}O_7$, 333.0974), implying nine degrees of unsaturation. The ^{13}C NMR spectrum showed resonances for two methoxyl, one methyl, one methylene, four methine, and nine quaternary carbons. The ^{1}H NMR spectrum displayed an aromatic proton δ_H [7.65 (s, H-9)], and two methoxyls at δ_H 3.99 (s, OCH$_3$-6)] and δ_H [3.95 (s, OCH$_3$-7)]. In comparison with the published data on bostrycin [5,6], both the ^{1}H NMR and ^{13}C NMR were similar, suggesting that compound **2** was a bostrycin derivative. Analysis of 1D NMR, ^{1}H–^{1}H COSY, heteronuclear single quantum correlation (HSQC), and HMBC data revealed the presence of one 1,2-dihydroxy-3-methylbutane unit and one pentasubstituted naphthoquinone moiety. In the HMBC spectrum, correlations of H-1 with C-9 and C-13, and of H-4 with C-10 and C-14, indicated 1,2-dihydroxy-3-methylbutane was connected to the naphthoquinone by linkage of C-1 with C-13 and of C-4 with C-14 (Figure 2). The phenolic OH was attached to C-10 by HMBC correlations of δ_H [12.18 (s, OH-10)] with C-10, C-11, and C-14. The aromatic proton δ_H [7.65 (s, H-9)] showed HMBC correlations with C-1, C-8, C-11, and C-14, suggesting C-9 was unsubstituted and the two methoxy groups were attached to C-6 and C-7. Therefore, the planar structure of compound **2** was identified as 3, 9-deoxy-7-methoxybostrycin and named aspetritone A (**2**).

The relative configuration of **2** was elucidated based on NOESY spectra (Figure 3). The strong NOESY correlations of H-1 with H-3 and of H-2 with H$_{ax}$-4 indicated that both H-1 and H-3 faced to the same side of the tetrahydrobenzene ring and H-2 oriented to the opposite side. Therefore, two possible isomers of (1*S*, 2*S*, 3*R*)-**2** and (1*R*, 2*R*, 3*S*)-**2** were proposed, and their ECD spectra were calculated by time-dependent density functional theory (TD-DFT). The experimental ECD spectrum of **2** was in good agreement with the calculated ECD spectrum of (1*S*, 2*S*, 3*R*)-**2** (Figure 4), and the axial–axial coupling constants of $^3J_{Hax\text{-}1,\,H\text{-}2}$ (7.03) and $^3J_{Hax\text{-}3,\,Hax\text{-}4}$ (11.46) indicated a half-chair form of the tetrahydrobenzene ring with all of OH-1, OH-2, and CH$_3$-3 in equatorial positions. In combining the NOESY correlations with the proton coupling constants, the absolute configuration of **2** was established as (1*S*, 2*S*, 3*R*)-3, 9-deoxy-7-methoxybostrycin (**2**).

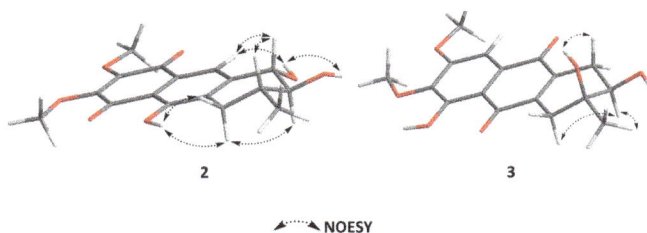

Figure 3. Key NOESY correlations of compounds **2** and **3**.

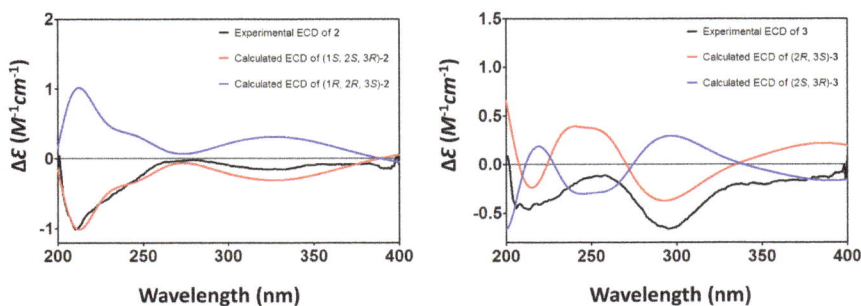

Figure 4. Calculated and Experimental ECD of compounds **2** and **3**.

Aspetritone B (**3**) was obtained as a yellow amorphous solid, and its molecular formula was determined as C$_{17}$H$_{18}$O$_7$ by HRESIMS (*m/z* 333.0979 [M − H]$^-$; calcd. for C$_{17}$H$_{17}$O$_7$,

333.0974), implying nine degrees of unsaturation. The ^{13}C NMR spectrum showed resonances for two methoxyl, one methyl, two methylene, two methine, and ten quaternary carbons. The ^{1}H NMR spectrum displayed an aromatic proton δ_H [7.16 (s, H-8)], and two methoxyls at δ_H [3.79 (s, OCH$_3$-6)] and δ_H [3.92 (s, OCH$_3$-7)]. In comparison with the published data of prisconnatanone A [7–9], both the ^{1}H NMR and ^{13}C NMR were similar, suggesting that compound **2** was a tetrahydroanthraquinone derivative. Analysis of 1D NMR, ^{1}H-^{1}H COSY, HSQC, and HMBC data revealed the presence of one 2,3-dihydroxy-3-methylbutane unit and one pentasubstituted naphthoquinone moiety. In HMBC spectra, correlations of H-1 with C-9 and C-14, and H-4 with C-10 and C-13, indicated that 2,3-dihydroxy-3-methylbutane was connected to the naphthoquinone by linkage of C-1 with C-13 and of C-4 with C-14. The phenolic OH was attached to C-5 by HMBC correlations of δ_H [12.09 (s, OH-5)] with C-5, C-6, and C-11. The aromatic proton δ_H [7.16 (s, H-8)] showed HMBC correlations with C-6, C-7, C-9, C-11, and C-12, suggesting C-8 was unsubstituted and the two methoxy groups were attached to C-6 and C-7 (Figure 2). Therefore, the planar structure of compound **3** was assigned as 1,2,3,4-tetrahydro-2,3,5-trihydroxy-3-methyl-6,7-dimethoxyanthracene-9,10-dione and named aspetritone B (**3**).

The relative configuration of **3** was elucidated based on NOESY spectra (Figure 3). The strong NOESY correlations of H$_{ax}$-1 with OH-3 and of H-2 with CH$_3$-3 indicated *cis*-configuration of OH-2 and OH-3. Therefore, two possible isomers of (2*R*, 3*S*)-**3** and (2*S*, 3*R*)-**3** were proposed, and their ECD spectra were calculated by TD-DFT. The experimental ECD spectrum of **3** was in good agreement with the calculated ECD spectrum of (2*R*, 3*S*)-**3** (Figure 4), and the axial–axial coupling constants of $^{3}J_{Hax-1, H-2}$ (11.63) indicated a half-chair form of the tetrahydrobenzene ring with OH-2 and CH$_3$-3 in equatorial positions. In combining the NOESY correlations with the proton coupling constants, the absolute configuration of **3** was established as (2*R*, 3*S*)-1,2,3,4-tetrahydro-2,3,5-trihydroxy-3-methyl-6,7-dimethoxyanthracene-9,10-dione (**3**).

The known compounds (**4–18**) were identified as 3,4-dimethyl-3″-prenylcandidusin A (**4**) [3], 4-methyl-3″-prenylcandidusin A (**5**) [3], 3,4-dimethyl-candidusin A (**6**) [3], candidusin A (**7**) [3], 4,4′-deoxy-terphenyllin (**8**) [10,11], 4″-deoxyterphenyllin (**9**) [10,11], 3-prenylterphenyllin (**10**) [10,11], terphenyllin (**11**) [12], 3-hydroxyterphenyllin (**12**) [13], 3-hydroxy-4″-deoxyterphenyllin (**13**) [11,14], 3″-prenylterphenyllin (**14**) [15], emodin (**15**) [16], 3-hydroxy-1,2,5,6-tetramethoxyanthracene-9,10-dione (**16**) [16], 3-hydroxy-2-hydroxymethyl-1-methoxyanthracene-9,10-dione (**17**) [16], and 1,2,3-trimethoxy-7-hydroxymethylanthracene-9,10-dione (**18**) [16] respectively, by comparing their spectroscopic data with those reported in the literature.

The cytotoxic and antimicrobial activity of compounds **1–18** was evaluated using cell lines of HeLa, A549, and Hep G2, and strains of methicillin-resistant *Staphylococcus aureus* (MRSA) (ATCC 43300, CGMCC 1.12409), *Vibrio vulnificus* MCCC E1758, *Vibrio rotiferianus* MCCC E385, and *Vibrio campbellii* MCCC E333 (Table 2). In comparison to the positive control chloramphenicol, for strains of MRSA, compounds **2** and **5** exhibited better antibacterial activities, and compounds **1**, **3**, **4**, **7**, **9–12**, and **14–16** showed weaker activities. Compound **5** displayed stronger anti-MRSA and lower cytotoxic activities than compound **2**. For strains of *Vibrio*, compound **5** showed stronger antibacterial activities than the other compounds, with MIC values ranging from 7–15 μg/mL. For cytotoxicity, compounds **2** and **10** showed significantly stronger activities than the other compounds with IC$_{50}$ values below 5 μM.

Table 2. Antibacterial and cytotoxic activities of compounds **1–18**. Data are expressed as mean ± SD values of three independent experiments, each made in triplicate.

Compound	MIC (µg/mL)					IC$_{50}$ (µM)		
	MRSA 1	MRSA 2	VV	VR	VC	HeLa	A549	Hep G2
1	31.33 ± 0.61	30.97 ± 0.78	31.47 ± 1.22	NA	15.10 ± 0.44	30.23 ± 1.32	24.53 ± 1.10	27.50 ± 1.57
2	7.53 ± 0.31	7.63 ± 0.21	15.61 ± 0.48	31.17 ± 0.35	15.53 ± 0.60	2.67 ± 0.60	3.13 ± 0.68	3.87 ± 0.74
3	15.27 ± 0.35	15.63 ± 0.45	15.47 ± 0.51	31.33 ± 0.23	15.77 ± 0.29	10.57 ± 0.93	4.67 ± 0.60	8.57 ± 0.83
4	15.67 ± 0.50	7.57 ± 0.73	15.58 ± 0.33	15.57 ± 0.30	NA	16.77 ± 0.45	21.07 ± 0.76	27.17 ± 0.29
5	3.80 ± 0.13	3.80 ± 0.22	7.77 ± 0.10	7.75 ± 0.18	15.57 ± 0.30	10.20 ± 0.50	13.07 ± 0.72	35.10 ± 1.00
6	NA	NA	NA	NA	NA	NA	NA	NA
7	31.47 ± 0.24	31.23 ± 0.10	31.42 ± 0.23	31.33 ± 0.19	NA	25.07 ± 0.81	19.07 ± 0.64	32.10 ± 2.00
8	NA	NA	NA	NA	NA	NA	NA	NA
9	31.30 ± 0.26	31.45 ± 0.22	31.37 ± 0.14	31.53 ± 0.31	31.47 ± 0.25	NA	NA	NA
10	15.53 ± 0.31	15.47 ± 0.23	31.43 ± 0.32	31.37 ± 0.21	NA	3.23 ± 0.40	3.87 ± 0.15	2.10 ± 0.20
11	31.47 ± 0.24	31.27 ± 0.16	31.37 ± 0.25	NA	NA	18.87 ± 1.27	12.33 ± 0.68	21.2 ± 0.35
12	31.30 ± 0.17	31.33 ± 0.12	31.43 ± 0.21	NA	NA	23.37 ± 0.84	36.07 ± 1.67	32.10 ± 2.65
13	NA	NA	NA	NA	NA	NA	NA	45.20 ± 1.00
14	31.33 ± 0.23	31.28 ± 0.10	31.25 ± 0.13	NA	31.43 ± 0.20	38.30 ± 1.50	NA	40.10 ± 0.90
15	15.65 ± 0.18	15.53 ± 0.12	15.73 ± 0.24	62.67 ± 0.15	31.35 ± 0.22	25.07 ± 0.81	22.17 ± 1.45	30.20 ± 0.87
16	31.32 ± 0.25	31.33 ± 0.23	NA	NA	NA	NA	NA	NA
17	NA	NA	NA	31.28 ± 0.14	NA	NA	45.63 ± 1.79	NA
18	NA	NA	NA	NA	NA	NA	NA	42.07 ± 1.07
erythromycin	NT	NT	1.92 ± 0.06	3.93 ± 0.03	7.68 ± 0.10	NT	NT	NT
chloramphenicol	7.67 ± 0.13	7.87 ± 0.08	NT	NT	NT	NT	NT	NT
doxorubicin	NT	NT	NT	NT	NT	0.50 ± 0.05	0.09 ± 0.01	1.06 ± 0.07

MRSA 1: methicillin-resistant *S. aureus* ATCC 43300; MRSA 2: methicillin-resistant *S. aureus* CGMCC 1.12409; VV: *V. vulnificus* MCCC E1758; VR: *V. rotiferianus* MCCC E385; VC: *V. campbellii* MCCC E333; NA: no activity at the concentration of 50 µg/mL (antibacterial) or 50 µM (cytotoxic); NT: not tested.

3. Materials and Methods

3.1. General Experimental Procedures

1D NMR and 2D NMR spectra were recorded on a Bruker DRX-400 instrument. HRESIMS was carried out on Bruker Daltonics Apex ultra 7.0 T Fourier transform mass spectrometer with an electrospray ionization source (Apollo II, Bruker Daltonics, Bremen, Germany). Optical rotations were measured with a P-1020 digital polarimeter (JASCO Corporation, Tokyo, Japan). CD spectra were measured on a J-715 spectropolarimeter (JASCO Corporation). The UV spectra were recorded on a UV-1800 spectrophotometer (Shimadzu, Japan). Thin-layer chromatography (TLC) plates (5 × 10 cm) were performed on GF254 (Branch of Qingdao Marine Chemical Co. Ltd., Qingdao, China) plates. For column chromatography (CC), RP-C18 (ODS-A, 50 µm, YMC, Kyoto, Japan), silica gel (200–300 mesh, 300–400 mesh, Branch of Qingdao Marine Chemical Co. Ltd., Qingdao, China), and Sephadex LH-20 (GE Healthcare Bio-Science AB, Pittsburgh, PA, USA) were used. The high performance liquid chromatography (HPLC) analysis was performed on a Waters 2695–2998 system (Waters, Milford, CT, USA). Semi-preparative HPLC was run with a P3000 pump (CXTH, Beijing, China) and a UV3000 ultraviolet-visible detector (CXTH, Beijing, China), using a preparative RP-C18 column (5 µm, 20 × 250 mm, YMC, Kyoto, Japan).

3.2. Fungal Material

Strain SP2-8-1 of A. tritici was isolated from the coral Galaxea fascicularis collected at Port Dickson, Malaysia, and was identified by ITS sequence homology (100% similarity with A. tritici CBS 266.81 with Genbank Accession No. KP987088.1 (max score 972, e value 0.0, query cover 100%)). The fungal strain was inoculated into a 15 mL centrifuge tube containing 3 mL of potato dextrose medium and cultured at 28 °C at 150 rpm for 3 days. Total genomic DNA was extracted as described by Lai et al. [17]. The internal transcribed spacer (ITS) region of rDNA was amplified by PCR using primers ITS1 (5′-TCCGTAGGTGAACCTGCGG-3′) and ITS4 (5′-TCCTCCGCTTATTGATATGC-3′). The PCR mixture consisted of 12.5 µL Taq premix (TaKaRa, Beijing, China), 0.25 µL (10 µM) of each primer, 0.75 µL dimethyl sulfoxide (DMSO), 10.25 µL dd H$_2$O, and 1 µL DNA template. After denaturation at 95 °C for 4 min, amplification was performed with 32 cycles of 30 s at 95 °C, 30 s at 55 °C, and 40 s at 72 °C, and a final extension at 72 °C for 7 min. The ITS1-5.8S-ITS2 rDNA sequence of the fungus has been submitted to GenBank with the accession number MF716581. A voucher specimen was deposited at the Third Institute of Oceanography, SOA, China. The working strain was prepared on potato dextrose agar slants and stored at 4 °C.

3.3. Fermentation, Extraction, and Isolation

Strain SP2-8-1 was cultured on PDA plates at 28 °C for 3 days. Then, six plugs (5 mm diameter) were transferred to 12 Erlenmeyer flasks (1 L), each containing 500 mL Czapek's medium (sucrose 30 g/L, NaNO$_3$ 3.0 g/L, MgSO$_4$·7H$_2$O 0.5 g/L, KH$_2$PO4 1.0 g/L, FeSO$_4$ 0.01 g/L and KCl 0.5 g/L) in sterile conditions. Erlenmeyer flasks were shaken on a rotary shaker at 28 °C and 120 rpm for 3 days to form seed cultures (1 × 10^8 spores/mL). Next, seed cultures (40 × 100 mL) were transferred to flasks (40 × 1 L) containing 45 g of millet and 105 g of rice per flask. After 28 days, the fermented culture was dried, smashed, and extracted with ethyl acetate (EtOAc). The EtOAc extract (220 g) was partitioned between petroleum ether (PE) and H$_2$O, and then between EtOAc and H$_2$O. Removal of the solvent of the EtOAc extract gave 150 g of residue, which was subject to silica gel (200–300 mesh) column chromatography, eluting with PE-EtOAC (9:1, 8.5:1.5, 8:2, 7.5:2.5, 6:4, 5:5, 4:6, *V:V*) to yield seven fractions, A–G. Further separation of fraction B (8.5:1.5, 24 g) was applied to silica gel column chromatography using PE-acetone and semi-preparative HPLC (80% methanol in H$_2$O, flow rate 12 mL/min) to give compounds **15** (16 mg), **18** (16 mg) and **16** (16 mg). Fraction C (8:2, 15 g) was further purified by semi-preparative HPLC (60% methanol in H$_2$O, flow rate 8 mL/min) and Sephadex LH-20 (50% chloroform in methanol) to give compounds **17** (16 mg) and **2** (4.5 mg). Fraction D (7.5:2.5, 10 g) was further purified by silica gel (200–300 mesh) column chromatography, eluting with hexane-EtOAc (6:4, *V:V*) and Sephadex LH-20 (methanol) to obtain compounds **3** (3.8 mg), **1** (32 mg) and **5** (16.9 mg). Fraction E (6:4, 10 g) was further separated by semi-preparative HPLC (80% methanol in H$_2$O, flow rate 8 mL/min), Sephadex LH-20 (85% methanol in H$_2$O), and preparative TLC to obtain compounds **4** (12 mg), **6** (15 mg), and **7** (20 mg). Fraction F (5:5, 20 g) was further separated by semi-preparative HPLC (80% methanol in H$_2$O, flow rate 8 mL/min) and Sephadex LH-20 (methanol) to obtain compounds **8** (24 mg), **9** (15 mg), and **12** (25 mg). Fraction G (4:6, 32 g) was further separated by semi-preparative HPLC (80% methanol in H$_2$O, flow rate 8 mL/min) and Sephadex LH-20 (methanol) to obtain compounds **11** (120 mg), **10** (60 mg), **13** (20 mg), and **14** (15 mg).

4-methyl-candidusin A (**1**): colorless amorphous solid; UV λ_{max} (methanol) nm (log ε): 295 (4.19); ^{1}H NMR and ^{13}C NMR data are shown in Table 1; HR-ESI-MS: m/z 367.11757 [M + H]$^+$ (Calcd. for 367.11816, C$_{21}$H$_{19}$O$_6$).

Aspetritone A (**2**): yellow amorphous solid; $[\alpha]_D^{20.0}$ − 350(c 0.15, MeOH); UV λ_{max} (methanol) nm (log ε): 257 (3.58); ^{1}H NMR and ^{13}C NMR data are shown in Table 1; HR-ESI-MS: m/z 333.0966 [M − H]$^-$ (Calcd. for 333.0974, C$_{17}$H$_{17}$O$_7$).

Aspetritone B (**3**): yellow amorphous solid; $[\alpha]_D^{20.0}$ − 156 (c 0.6, MeOH); UV λ_{max} (methanol) nm (log ε): 265 (3.73); 283 (3.53); ^{1}H NMR and ^{13}C NMR data are shown in Table 1; HR-ESI-MS: m/z 333.0979 [M − H]$^-$ (Calcd. for 333.0974, C$_{17}$H$_{17}$O$_7$).

3.4. Antibacterial Assay

Antibacterial activities against MRSA (ATCC 43300, CGMCC 1.12409), *V. rotiferianus* (MCCC E385), *V. vulnificus* (MCCC E1758), and *V. campbellii* (MCCC E333) were tested by continuous dilution in 96-well plates using resazurin as a surrogate indicator. Blue resazurin was reduced by metabolically active bacteria to pink resorufin. A mid-logarithmic-phase tested strain was added at a starting inoculum of 5 × 10^5 CFU/mL to the plate containing tested compound (final concentration ranging from 250 to 0.98 μg/mL in two-fold dilution) plus 10% resazurin solution (6.75 mg/mL in sterile water). The foil covered plate was incubated for 24 h with shaking at 37 °C. After that, by observing the blue-to-pink color change, the MIC value was determined to be the lowest concentration that did not induce the color change [18–20].

3.5. Cytotoxicity Assay

Hela (cervical cancer cell), Hep G2 (human liver cancer cell), and A549 (adenocarcinomic human alveolar basal epithelial cell) cells were maintained in DMEM, MEM, and F-12K medium respectively, and supplied with 10% FBS, 100 U/mL of penicillin, and 100 mg/mL of streptomycin [21]. Cells were grown in a humidified chamber with 5% CO_2 at 37 °C. For cytotoxicity assays, cells were seeded at a density of 5000 cells per well in 96-well plates, grown at 37 °C for 12 h, and then treated with tested compound at five different concentrations (100 µL medium/well). The cytotoxicity was measured by Cell Counting Kit-8 (CCK-8) (DOJINDO) at 48 h post-treatment, following the manufacturer's instructions.

CCK-8 assay is based on the conversion of a tetrazolium salt, 2-(2-methoxy-4-nitrophenyl)-3-(4-nitrophenyl)-5-(2,4-disulfophenyl)-2*H*-tetrazolium, monosodium salt (WST-8), and a water-soluble formazan dye, upon reduction by dehydrogenases in the presence of an electronmediator [22]. WST-8 is reduced by dehydrogenases in cells to give an orange colored product (formazan). The amount of the formazan dye is directly proportional to the number of living cells.

In brief, 10 µL of CCK-8 solution was added to each well of the 96-well plates. After incubation at 37 °C for 2 h, the absorbance at 450 nm was measured using a SpectraMAX M5 microplate reader. Wells with only culture medium and CCK-8 solution were used to determine the background, and cells treated with DMSO were included as the negative controls [21].

3.6. ECD Calculation

Conformational analysis was initially performed using Confab [23] with the MMFF94 force field for all configurations. Room-temperature equilibrium populations were calculated according to the Boltzmann distribution law Equation (1). The conformers with Boltzmann-populations of over 1% were chosen for ECD calculations. The energies and populations of all dominative conformers were provided in Table S1.

$$\frac{N_i}{N} = \frac{g_i e^{-\frac{E_i}{k_B T}}}{\sum g_i e^{-\frac{E_i}{k_B T}}} \tag{1}$$

N_i is the number of conformer i with energy E_i and degeneracy g_i at temperature T, and k_B is the Boltzmann constant.

The theoretical calculation was carried out using Gaussian 09. First, the chosen conformer was optimized at PM6 using the semi-empirical theory method, and then optimized at B3LYP/6-311G** in methanol using the conductor-like polarizable continuum model (CPCM) (Table S2). The theoretical calculation of ECD was conducted in methanol using TD-DFT at the same theory level. Rotatory strengths for a total of 50 excited states were calculated. The ECD spectrum is simulated in SpecDis [24] by overlapping Gaussian functions for each transition.

4. Conclusions

In current research, we have isolated three novel compounds, 4-methyl candidusin A (**1**), aspetritone A (**2**), and aspetritone B (**3**), together with two prenylcandidusin derivatives (**4–5**), two candidusin derivatives (**6–7**), seven terphenyllin derivatives (**8–14**), and four anthraquinone derivatives (**15–18**). Candidusin can be deduced to be a cyclization product of terphenyllin between C-6 and C-2′ via an oxygen atom. C-prenylation plays an important role in diversification of natural compounds, especially for flavonoids and coumarins [25]. These compounds exhibit significant in vitro biological activities (cytotoxic, antibacterial, osteogenic, antioxidant, and anti-inflammatory activities) [25]. Prenylation in polyhydroxy-p-terphenyl analogues, as 3,4-dimethyl-3″-prenylcandidusin A (**4**), 4-methyl-3″-prenylcandidusin A (**5**),

3-prenylterphenyllin (**10**), and 3″-prenylterphenyllin (**14**), described in this paper, preliminarily influences the cytotoxicity and antibacterial activities, compared to the un-prenylated terphenyllin and candidusin derivatives. Chemical structures of compounds **2**, **3**, and **15–18** also preliminarily indicate that the special tetrahydrobenzene moiety located in compounds **2** and **3** attributed to their relatively strong bioactivity. Therefore, isolation or synthesis of more prenylated-terphenyllin, prenylated-candidusin, and tetrahydroanthraquinone derivatives deserves attention as an important aspect of structure–activity relationship studies.

Supplementary Materials: The following are available online at www.mdpi.com/1660-3397/15/11/348/s1: NMR spectra for compounds **1–3** as well as computational data for compounds **1–2**.

Acknowledgments: This research was supported by the International Science & Technology Cooperation Program of China [2015DFA20500]. We would like to thank all of the members of Peng Cai's group in the Institute of Urban Environment, Chinese Academy of Sciences, as well as Wei Xu in the Third institute of Oceanography, SOA, for their assistance in this work.

Author Contributions: Weiyi Wang carried out the isolation and structural elucidation, and wrote this paper. Yanyan Liao performed the antibacterial and cytotoxic activity evaluations. Chao Tang and Xiaomei Huang carried out the fermentation of fungi. Zhuhua Luo contributed to this work by ITS sequencing. Jianming Chen and Peng Cai conceived and designed the experiments. Peng Cai contributed to the revision of the paper.

Conflicts of Interest: The authors declare no conflict of interest.

References

1. Fouillaud, M.; Venkatachalam, M.; Girard-Valenciennes, E.; Caro, Y.; Dufosse, L. Anthraquinones and derivatives from marine-derived fungi: Structural diversity and selected biological activities. *Mar. Drugs* **2016**, *14*, 64. [CrossRef] [PubMed]

2. Schueffler, A.; Anke, T. Fungal natural products in research and development. *Nat. Prod. Rep.* **2014**, *31*, 1425–1448. [CrossRef] [PubMed]

3. Cai, S.; Sun, S.; Zhou, H.; Kong, X.; Zhu, T.; Li, D.; Gu, Q. Prenylated polyhydroxy-p-terphenyls from *Aspergillus taichungensis* ZHN-7-07. *J. Nat. Prod.* **2011**, *74*, 1106–1110. [CrossRef] [PubMed]

4. Liu, S.S.; Zhao, B.B.; Lu, C.H.; Huang, J.J.; Shen, Y.M. Two new p-terphenyl derivatives from the marine fungal strain *Aspergillus* sp. AF119. *Nat. Prod. Commun.* **2012**, *7*, 1057–1062. Available online: https://www.ncbi.nlm.nih.gov/pubmed/22978228 (accessed on 1 August 2012). [PubMed]

5. Xu, J.; Nakazawa, T.; Ukai, K.; Kobayashi, H.; Mangindaan, R.E.P.; Wewengkang, D.S.; Rotinsulu, H.; Namikoshi, M. Tetrahydrobostrycin and 1-Deoxytetrahydrobostrycin, Two New Hexahydroanthrone Derivatives, from a Marine-derived Fungus *Aspergillus* sp. *J. Antibiot.* **2008**, *61*, 415–419. [CrossRef] [PubMed]

6. Chen, H.; Zhong, L.; Long, Y.; Li, J.; Wu, J.; Liu, L.; Chen, S.; Lin, Y.; Li, M.; Zhu, X.; et al. Studies on the synthesis of derivatives of marine-derived bostrycin and their structure-activity relationship against tumor cells. *Mar. Drugs* **2012**, *10*, 932–952. [CrossRef] [PubMed]

7. Wang, C.; Ding, X.; Feng, S.X.; Guan, Q.; Zhang, X.P.; Du, C.; Di, Y.T.; Chen, T. Seven new tetrahydroanthraquinones from the root of *Prismatomeris connata* and their cytotoxicity against lung tumor cell growth. *Molecules* **2015**, *20*, 22565–22577. [CrossRef] [PubMed]

8. Ondeyka, J.; Buevich, A.V.; Williamson, R.T.; Zink, D.L.; Polishook, J.D.; Occi, J.; Vicente, F.; Basilio, A.; Bills, G.F.; Donald, R.G.; et al. Isolation, structure elucidation, and biological activity of altersolanol P using *Staphylococcus aureus* fitness test based genome-wide screening. *J. Nat. Prod.* **2014**, *77*, 497–502. [CrossRef] [PubMed]

9. Debbab, A.; Aly, A.H.; Edrada-Ebel, R.; Wray, V.; Muller, W.E.; Totzke, F.; Zirrgiebel, U.; Schachtele, C.; Kubbutat, M.H.; Lin, W.H.; et al. Bioactive metabolites from the endophytic fungus *Stemphylium globuliferum* isolated from *Mentha pulegium*. *J. Nat. Prod.* **2009**, *72*, 626–631. [CrossRef] [PubMed]

10. Guo, Z.K.; Yan, T.; Guo, Y.; Song, Y.C.; Jiao, R.H.; Tan, R.X.; Ge, H.M. P-Terphenyl and diterpenoid metabolites from endophytic *Aspergillus* sp. YXf3. *J. Nat. Prod.* **2012**, *75*, 15–21. [CrossRef] [PubMed]

11. Huang, H.; Feng, X.; Xiao, Z.; Liu, L.; Li, H.; Ma, L.; Lu, Y.; Ju, J.; She, Z.; Lin, Y. Azaphilones and p-terphenyls from the mangrove endophytic fungus *Penicillium chermesinum* (ZH4-E2) isolated from the South China Sea. *J. Nat. Prod.* **2011**, *74*, 997–1002. [CrossRef] [PubMed]

12. Marchelli, R.; Vining, L.C. Terphenyllin, A novel p-terphenyl metabolite from *Aspergillus candidus*. *J. Antibiot. (Tokyo)* **1975**, *28*, 328–331. [CrossRef] [PubMed]

13. Kurobane, I.; Vining, L.C.; McInnes, A.G.; Smith, D.G. 3-Hydroxyterphenyllin, a new metabolite of *Aspergillus candidus* Structure elucidation by H and C nuclear magnetic resonance spectroscopy. *J. Antibiot. (Tokyo)* **1979**, *32*, 559–564. [CrossRef] [PubMed]

14. Kamigauchi, T.; Sakazaki, R.; Nagashima, K.; Kawamura, Y.; Yasuda, Y.; Matsushima, K.; Tani, H.; Takahashi, Y.; Ishii, K.; Suzuki, R.; et al. Terprenins, Novel immunosuppressants produced by *Aspergillus candidus*. *J. Antibiot. (Tokyo)* **1998**, *51*, 445–450. [CrossRef]

15. Zhang, W.; Wei, W.; Shi, J.; Chen, C.; Zhao, G.; Jiao, R.; Tan, R. Natural phenolic metabolites from endophytic *Aspergillus* sp. IFB-YXS with antimicrobial activity. *BioOrg. Med. Chem. Lett.* **2015**, *25*, 2698–2701. [CrossRef] [PubMed]

16. Feng, S.-X.; Hao, J.; Chen, T.; Qiu, S.X. A New Anthraquinone and two new tetrahydroanthraquinones from the Roots of *Prismatomeris connata*. *Helv. Chim. Acta* **2011**, *94*, 1843–1849. [CrossRef]

17. Lai, X.; Cao, L.; Tan, H.; Fang, S.; Huang, Y.; Zhou, S. Fungal communities from methane hydrate-bearing deep-sea marine sediments in South China Sea. *ISME J.* **2007**, *1*, 756–762. [CrossRef] [PubMed]

18. Chhillar, A.K.; Gahlaut, A. Evaluation of antibacterial potential of plant extracts using resazurin based microtiter dilution assay. *Int. J. Pharm. Pharm. Sci.* **2013**, *5*, 372–376.

19. Wibowo, A.; Ahmat, N.; Hamzah, A.S.; Low, A.L.; Mohamad, S.A.; Khong, H.Y.; Sufian, A.S.; Manshoor, N.; Takayama, H.; Malaysianol, B. An oligostilbenoid derivative from *Dryobalanops lanceolata*. *Fitoterapia* **2012**, *83*, 1569–1575. [CrossRef] [PubMed]

20. Coban, A.Y. Rapid determination of methicillin resistance among *Staphylococcus aureus* clinical isolates by colorimetric methods. *J. Clin. Microbiol.* **2012**, *50*, 2191–2193. [CrossRef] [PubMed]

21. Han, S.B.; Shin, Y.J.; Hyon, J.Y.; Wee, W.R. Cytotoxicity of voriconazole on cultured human corneal endothelial cells. *Antimicrob. Agents. Chemother.* **2011**, *55*, 4519–4523. [CrossRef] [PubMed]

22. Ishiyama, M.; Tominaga, H.; Shiga, M.; Sasamoto, K.; Ohkura, Y.; Ueno, K. A combined assay of cell viability and in vitro cytotoxicity with a highly water-soluble tetrazolium salt, neutral red and crystal violet. *Biol. Pharm. Bull.* **1996**, *19*, 1518–1520. [CrossRef] [PubMed]

23. O'Boyle, N.M.; Vandermeersch, T.; Flynn, C.J.; Maguire, A.R.; Hutchison, G.R. Confab-Systematic generation of diverse low-energy conformers. *J. Cheminform.* **2011**, *3*, 8. [CrossRef] [PubMed]

24. Bruhn, T.; Schaumloffel, A.; Hemberger, Y.; Bringmann, G. SpecDis: Quantifying the comparison of calculated and experimental electronic circular dichroism spectra. *Chirality* **2013**, *25*, 243–249. [CrossRef] [PubMed]

25. Chen, X.; Mukwaya, E.; Wong, M.S.; Zhang, Y. A systematic review on biological activities of prenylated flavonoids. *Pharm. Biol.* **2014**, *52*, 655–660. [CrossRef] [PubMed]

marine drugs

MDPI

Article

Asperindoles A–D and a *p*-Terphenyl Derivative from the Ascidian-Derived Fungus *Aspergillus* sp. KMM 4676

Elena V. Ivanets [1], Anton N. Yurchenko [1,*], Olga F. Smetanina [1], Anton B. Rasin [1],
Olesya I. Zhuravleva [1,2], Mikhail V. Pivkin [1], Roman S. Popov [1], Gunhild von Amsberg [3],
Shamil Sh. Afiyatullov [1] and Sergey A. Dyshlovoy [1,2,3]

[1] G.B. Elyakov Pacific Institute of Bioorganic Chemistry, Far Eastern Branch of the Russian Academy of
    Sciences, Prospect 100-letiya Vladivostoka, 159, Vladivostok 690022, Russia; ev.ivanets@yandex.ru (E.V.I.);
    smetof@rambler.ru (O.F.S.); abrus__54@mail.ru (A.B.R.); zhuravleva.oi@dvfu.ru (O.I.Z.);
    oid27@mail.ru (M.V.P.); prs_90@mail.ru (R.S.P.); afiyat@piboc.dvo.ru (S.S.A.); dyshlovoy@gmail.com (S.A.D.)
[2] School of Natural Science, Far Eastern Federal University, Sukhanova St., 8, Vladivostok 690000, Russia
[3] Laboratory of Experimental Oncology, Department of Oncology, Hematology and Bone Marrow
    Transplantation with Section Pneumology, Hubertus Wald-Tumorzentrum, University Medical Center
    Hamburg-Eppendorf, 20246 Hamburg, Germany; g.von-amsberg@uke.de
*   Correspondence: yurchant@ya.ru; Tel.: +7-423-231-1168

Received: 7 June 2018; Accepted: 3 July 2018; Published: 9 July 2018

Abstract: Four new indole-diterpene alkaloids asperindoles A–D (**1–4**) and the known *p*-terphenyl
derivative 3″-hydroxyterphenyllin (**5**) were isolated from the marine-derived strain of the fungus
Aspergillus sp., associated with an unidentified colonial ascidian. The structures of **1–5** were
established by 2D NMR and HRESIMS data. The absolute configurations of all stereocenters
of **1–4** were determined by the combination of ROESY data, coupling constants analysis, and
biogenetic considerations. Asperindoles C and D contain a 2-hydroxyisobutyric acid (2-HIBA)
residue, rarely found in natural compounds. Asperindole A exhibits cytotoxic activity against
hormone therapy-resistant PC-3 and 22Rv1, as well as hormone therapy-sensitive human prostate
cancer cells, and induces apoptosis in these cells at low-micromolar concentrations.

Keywords: marine-derived fungi; secondary metabolites; indole-diterpenoids; cytotoxicity

---

1. Introduction

Marine fungi are promising and prolific sources of new biological active compounds.
Fungi of the genus *Aspergillus*, section *Candidi* (*A. candidus*, *A. campestris*, *A. taichungensis*,
A. tritici), are known to produce several types of *p*-terphenyl derivatives, such as terphenyllins
(terphenyllin [1], 3-hydroxyterphenyllin [2], terprenins [3]) and candidusins (candidusins A–C [4],
prenylcandidusins A–C [5]), and a number of flavonoid derivatives (e.g., chlorflavonin [6],
chlorflavonin A [7]). These compounds exhibit antioxidant [8,9], cytotoxic [5,9,10], antimicrobial [9],
and immunosuppressive activities [9]. Recently, several indole and pyrrolidine alkaloids unusual for
this fungal group were isolated from a Thai sponge-derived *A. candidus* strain [11].

Indole-diterpene alkaloids are widely represented among the fungal metabolites.
These compounds have been isolated from fungi of the genera *Claviceps*, *Acremonium*, *Eupenicillium*,
Penicillium, and *Aspergillus* (including *Emericella striata*). Most natural indole-diterpenes have an
invariable framework (Figure 1). Usually, C-7, C-13, and C-27 are oxidized. The oxygenation of
C-27 is often followed by the formation of an ether bridge between C-27 and C-7 with inversion of
the stereoconfiguration at C-7 [12]. Interestingly, some fungi produce metabolites with relatively
rare features in the classical indole-diterpene backbone. For example, *Acremonium lolii* produces

---

indoles diprenylated at C-20 and C-21, together with oxygenated derivatives [13]. Alkaloids with a 1,3-dioxane moiety joined at C-9 and C-10 with the F-ring from *A. lolii* have also been reported [14,15]. Many of such compounds showed tremorgenic [16], cytotoxic [17,18], and antiinsectan [19] activities, and some of them are antagonists of cannabinoid receptors [20].

Based on promising screening results in search of producers of biologically active compounds, the marine-derived fungus *Aspergillus* sp. KMM 4676, which is associated with an unidentified colonial ascidian (from the Shikotan Island in the Pacific Ocean), was selected for further studies. During earlier examinations of this fungal strain, five known *p*-terphenyls and one known flavonoid were isolated [21]. Herein, we describe the results of subsequent comprehensive chemical and bioactivity investigations of the extracts of strain KMM 4676, leading to the characterization of four new natural compounds.

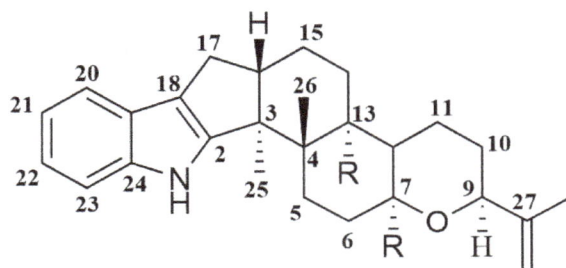

Figure 1. Usual framework of indole-diterpenes.

2. Results

The HRESIMS spectrum of **1** exhibited a pseudo-molecular peak at m/z 526.1980 [M + H]$^+$, showing the characteristic isotope pattern with one chlorine atom, therefore establishing its molecular formula as $C_{29}H_{32}NO_6Cl$, which was supported by the ^{13}C NMR spectrum.

Inspection of the ^1H and ^{13}C NMR data (Table 1, Figures S1–S2) of **1** revealed the presence of three quaternary methyls (δ_C 16.1, 17.0, 23.4; δ_H 1.02, 1.21, 1.31), one acetate methyl (δ_C 21.8, δ_H 2.07), six methylene sp^3 (δ_C 20.8, 26.3, 26.9, 30.0, 31.5, 64.8; δ_H 1.66, 1.77, 1.91, 1.93, 2.00, 2.11, 2.30, 2.40, 2.55, 2.60, 3.67, 4.04), two methine sp^3 (δ_C 48.3, 78.3, δ_H 2.72, 4.74), four methine sp^2 (δ_C 111.3, 118.6, 118.7, 119.9; δ_H 6.11, 6.91, 7.26, 7.27), three quaternary oxygen-bearing sp^3 (δ_C 75.0, 77.0, 93.6), two quaternary sp^3 (δ_C 38.5, 51.4), and eight quaternary sp^2 (δ_C 115.0, 123.3, 123.7, 140.2, 154.0, 159.1, 170.2, 195.9) carbons, as well as a NH singlet (δ_H 10.73) and an OH singlet (δ_H 5.10).

The ^1H and ^{13}C NMR spectra of **1** (Table 1) resembled those of paspalinine [16], suggesting that **1** has an indole-diterpene core similar to that of paspalinine. However, the differences in chemical shift values of C-19 (δ_C 123.3) and C-22 (δ_C 123.7) of **1** from the corresponding carbons in paspalinine [16]; the HMBC correlations (Figure 2, Figure S5) from H-20 (δ_H 7.26) to C-18 (δ_C 115.0) and C-22, from H-21 (δ_H 7.26) to C-19 and C-23 (δ_C 111.3), and from H-23 (δ_H 7.26) to C-19 and C-21 (δ_C 118.7); and the coupling constants $J_{H20-H21}$ (8.5 Hz) and $J_{H21-H23}$ (2.1 Hz) suggested the presence of a chlorine atom on C-22 of **1**.

Table 1. ^{13}C NMR data (125 MHz, δ in ppm, DMSO-d_6) for asperindoles A–D (**1–4**).

Position	1	2	3	4
2	154.0, C	152.8, C	154.1, C	152.8, C
3	51.4, C	51.2, C	51.4, C	51.2, C
4	38.5, C	38.6, C	38.5, C	38.6, C
5	26.3, CH$_2$	26.2, CH$_2$	26.3, CH$_2$	26.2, CH$_2$
6	30.0, CH$_2$	30.1, CH$_2$	30.0, CH$_2$	30.0, CH$_2$
7	93.6, C	93.6, C	93.6, C	93.5, C
9	78.3, CH	78.3, CH	78.3, CH	78.6, CH
10	195.9, C	195.9, C	195.9, C	195.7, C
11	119.9, CH	119.9, CH	119.9, CH	119.8, CH
12	159.1, C	159.2, C	159.1, C	159.2, C
13	77.0, C	77.0, C	77.0, C	77.0, C
14	31.5, CH$_2$	31.6, CH$_2$	31.5, CH$_2$	31.5, CH$_2$
15	20.8, CH$_2$	20.9, CH$_2$	20.82, CH$_2$	20.9, CH$_2$
16	48.3, CH	48.3, CH	48.3, CH	48.4, CH
17	26.9, CH$_2$	27.1, CH$_2$	26.9, CH$_2$	27.1, CH$_2$
18	115.0, C	114.8, C	115.0, C	114.8, C
19	123.3, C	124.6, C	123.3, C	124.6, C
20	118.6, CH	117.5, CH	118.6, CH	117.5, CH
21	118.7, CH	118.3, CH	118.7, CH	118.3, CH
22	123.7, C	119.1, CH	123.7, C	119.1, CH
23	111.3, CH	111.8, CH	111.3, CH	111.8, CH
24	140.2, C	139.9, C	140.2, C	139.9, C
25	16.1, CH$_3$	16.2, CH$_3$	16.1, CH$_3$	16.2, CH$_3$
26	23.4, CH$_3$	23.4, CH$_3$	23.4, CH$_3$	23.3, CH$_3$
27	75.0, C	75.0, C	75.8, C	75.8, C
28	64.8, CH$_2$	64.8, CH$_2$	64.1, CH$_2$	64.1, CH$_2$
29	17.0, CH$_3$	17.0, CH$_3$	16.4, CH$_3$	16.4, CH$_3$
1′	170.2, C	170.2, C	171.1, C	171.1, C
2′	21.8, CH$_3$	21.8, CH$_3$	77.9, C	77.9, C
3′			23.9, CH$_3$	23.9, CH$_3$
4′			24.2, CH$_3$	24.2, CH$_3$
1″			169.3, C	169.3, C
2″			20.75, CH$_3$	20.8, CH$_3$

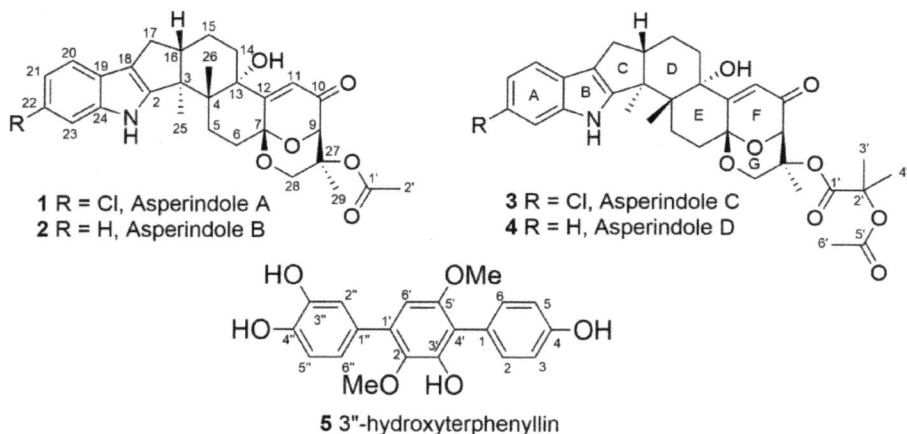

1 R = Cl, Asperindole A
2 R = H, Asperindole B

3 R = Cl, Asperindole C
4 R = H, Asperindole D

5 3″-hydroxyterphenyllin

Figure 2. Chemical structures of **1–5**.

The HMBC correlations (Figure 3, Figure S5) from H-28β (δ_H 3.67) to C-27 (δ_C 75.0), C-29 (δ_C 17.0), and C-1′ (δ_C 170.2); from H-9 (δ_H 4.74) to C-7 (δ_C 93.6), C-28 (δ_C 64.8), and C-27; from H-28α (δ_H 4.04) to C-7, C-9 (δ_C 78.3), and C-27; and from H$_3$-29 (δ_H 1.21) to C-27 suggested the presence of a 1,3-dioxane ring with an acetoxy group at C-27. The W-type coupling constant $J_{H9-H28\alpha}$ (2.5 Hz) and ROESY correlations (Figure 4, Figure S6) of H-28β with H-11 (δ_H 6.11), H$_3$-29, H$_3$-26 (δ_H 1.02), and of H-28α with H$_3$-29 indicated a relative configuration of chiral centers in the 1,3-dioxane ring as 7R*, 9R*, 27S*. The ROESY correlations (Figure 4, Figure S6) of H$_3$-25 (δ_H 1.31) with H-5α (δ_H 1.93), 13-OH (δ_H 5.10), H-6α (δ_H 2.00), and H-15α (δ_H 1.91), and of H-16 (δ_H 2.72) with H$_3$-26 (δ_H 1.02) suggested the relative configurations of the stereogenic carbons of the C–G rings in **1** as 3S*, 4R*, 13S*, 16R*. The absolute configurations of all stereocentres in **1** was proposed as 3S, 4R, 7R, 9R, 13S, 16R, 27S—the same as those in paspalinine—based on biosynthetic considerations, and was confirmed by the comparison of the ECD (electronic circular dichroism) spectral data with that of paspalinine [17] (Figure 5). Compound **1** was named asperindole A. It should be noted that chlorinated indolediterpenes are rare in nature [12,22–25].

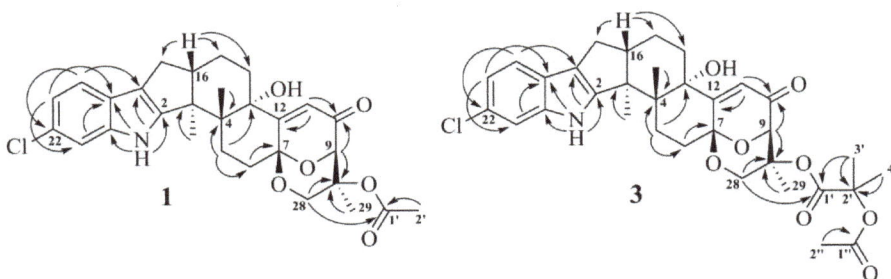

Figure 3. Key HMBC correlations of **1** and **3**.

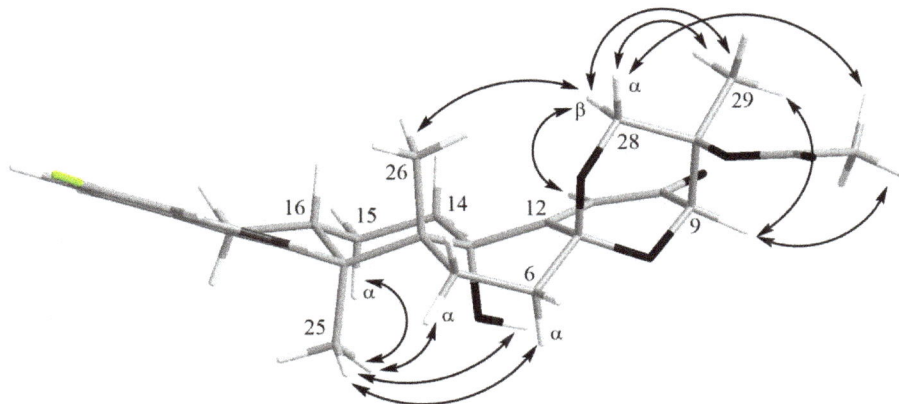

Figure 4. Key ROESY correlations in asperindole A (**1**).

The molecular formula of **2** was determined as $C_{29}H_{33}NO_6$ by a HRESIMS peak at *m/z* 490.2188 [M − H]$^-$, which was supported by the ^{13}C NMR spectrum. The general features of the ^{1}H and ^{13}C NMR spectra (Table 1, Figures S7 and S8) of **2** resemble those of **1**, with the exception of the proton and carbon signals of an indole moiety, as well as the absence of a chlorine atom as evidenced by the HRESIMS spectrum. The coupling constants and the multiplicity of the aromatic protons in ring A (H-20, δ_H 7.25, d, *J* = 7.6 Hz; H-21, δ_H 6.89, t, *J* = 7.6 Hz; H-22, δ_H 6.93, t, *J* = 7.6 Hz; and H-23, δ_H 7.27, d, *J* = 7.6 Hz) allowed the conclusion to be made that **2** is a nonchlorinated analogue of **1**. Compound **2** was therefore named asperindole B.

The molecular formula of **3** was established as $C_{33}H_{38}NO_8Cl$ on the basis of the HRESIMS, containing a peak at *m/z* 610.2206 [M − H]⁻, and was supported by the ¹³C NMR spectrum. The analysis of the NMR data (Figures S14–S20) for **3** revealed the presence of the same indole-diterpene framework as that in **1**, with the exception of the proton and carbon signals in a 1,3-dioxane ring, as well as the presence of two methyl (δ_C 23.9, 24.2), an ester carbonyl (δ_C 171.1), and an oxygen-bearing quaternary *sp*³ (δ_C 77.9) carbons. The HMBC correlations (Figure 3, Figure S19) from H-3' (δ_H 1.52) and H-4' (δ_H 1.54) to C-2' (δ_C 77.9), and from H-6' (δ_H 2.04) to C-5' (δ_C 169.3) suggested the presence of an acetylated residue of 2-hydroxyisobutyric acid (2-HIBA) in **3**. This was corroborated by the molecular weight of **3**, which was 86 amu ($C_4H_6O_2$) greater than that of **1**. The ROESY correlations of **3** (Figure S20) were similar to those in **1** (Figure 4, Figure S6). Based on these data and together with the ECD spectrum of **1** (Figure 5), the absolute configurations of all stereocentres in **3** were proposed to be the same as those in asperindole A. Consequently, **3** was named asperindole C. To the best of our knowledge, the 2-HIBA residue is unique amongst naturally occurring compounds.

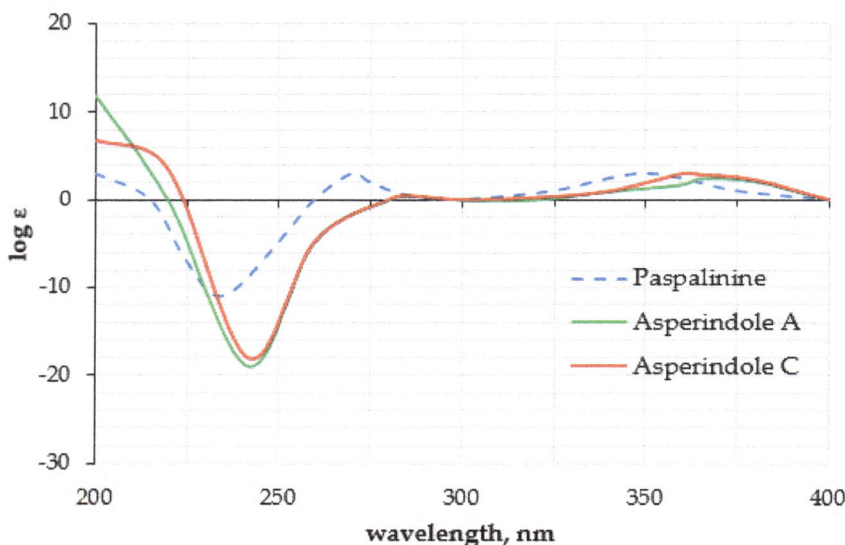

Figure 5. Experimental ECD data of **1**, **3**, and paspalinine [17].

The HRESIMS spectrum of **4** exhibited the [M − H]⁻ peak at *m/z* 576.2594, corresponding to $C_{33}H_{39}NO_8$, which was supported by the ¹³C NMR spectrum. The general features of the ¹H and ¹³C NMR spectra (Table 2, Figures S21 and S22) of **4** resembled those of **3**, with the exception of some proton and carbon signals of the indole moiety. Similar to **2**, the coupling constants and multiplicity of the aromatic protons in ring A (H-20, δ_H 7.25, d, *J* = 7.5 Hz; H-21, δ_H 6.88, brt, *J* = 7.2 Hz; H-22, δ_H 6.92, brt, *J* = 7.1 Hz; and H-23, δ_H 7.27, d, *J* = 6.9 Hz) led to the conclusion that **4** is a nonchlorinated analogue of **3**. Compound **4** was therefore named asperindole D.

Table 2. ^{1}H NMR data (δ in ppm, *J* in Hz, DMSO-d_6) for asperindoles A–D (**1–4**).

Position	1 *	2 **	3 **	4 **
NH	10.73, brs	10.54, brs	10.73, s	10.52, s
5α	1.93, m	1.96, m	1.95, m	1.96, m
5β	2.40, dd (13.4, 10.0)	2.41, t (12.3)	2.39, dd (13.4,10.0)	2.41, t (12.3)
6α	2.00, dd (12.9, 8.6)	1.99, m	1.95, dd (12.9, 8.6)	1.95, dd (12.9, 8.6)
6β	2.55, m	2.55, m	2.55, m	2.55, m
9	4.74, d (2.3)	4.74, d (2.1)	4.74, d (2.3)	4.63, d (2.4)
11	6.11, s	6.11, s	6.11, s	6.12, s
14α	2.11, dt (13.6, 2.8)	2.12, brd (13.4)	2.12, dt (13.6, 2.8)	
14β	1.77, td (13.2, 4.5)	1.78, brt (13.4)	1.76, td (13.2, 4.5)	
15α	1.91, m	1.91, m	1.91, m	1.91, m
15β	1.66, m	1.66, m	1.65, m	1.65, m
16	2.72, m	2.72, m	2.72, m	2.63, m
17α	2.30, dd (13.0, 10.9)	2.30, t (12.3)	2.30, dd (13.0, 10.9)	2.31, dd (13.0, 10.9)
17β	2.60, dd (13.0, 6.4)	2.60, dd (12.3, 6.6)	2.60, dd (13.0, 6.4)	2.60, dd (13.0, 6.2)
20	7.26, d (8.6)	7.25, d (7.6)	7.26, d (8.3)	7.25, d (7.5)
21	6.91, dd (8.3, 1.9)	6.89, t (7.6)	6.91, dd (8.3, 2.0)	6.88, brt (7.2)
22		6.93, t (7.6)		6.92, brt (7.1)
23	7.27, d (2.2)	7.27, d (7.6)	7.25, d (2.0)	7.27, d (6.9)
25	1.31, s	1.30, s	1.31, s	1.30, s
26	1.02, s	1.03, s	1.02, s	1.03, s
28α	4.04, dd (13.4, 2.5)	4.05, dd (13.3, 2.1)	4.11, dd (13.2, 2.5)	4.11, dd (13.4, 2.4)
28β	3.67, d (13.4)	3.68, d (13.3)	3.68, d (13.2)	3.69, d (13.4)
29	1.21, s	1.21, s	1.17, s	1.18, s
3$'$			1.52, s	1.52, s
4$'$			1.54, s	1.54, s
2$''$	2.07, s	2.07, s	2.04, s	2.04, s
13-OH	5.10, s	5.08, s	5.11, s	5.08, s

^{1}H NMR spectroscopic data were measured at * 700 MHz and ** 500 MHz, respectively.

The molecular formula of **5** was determined as $C_{20}H_{18}O_6$, based on a pseudo-molecular peak at *m/z* 353.1013 [M − H]$^-$ from the HRESIMS spectrum. This was supported by the ^{13}C NMR spectrum. A close inspection of the ^{1}H and ^{13}C NMR data (Table 3, Figures S23 and S24) of **5** revealed the presence of eight aromatic protons (δ_H 6.47, 6.85, 2H; 6.91, 7.02, 7.19, 7.25, 2H) and eight methine sp^2 (δ_C 104.8, 115.9, 115.9, 116.8, 117.5, 122.0, 133.7, 133.7), six oxygen-bearing quaternary sp^2 (δ_C 140.8, 146.3, 146.4, 149.8, 155.1, 157.6), and four quaternary sp^2 (δ_C 118.3, 126.8, 131.9, 134.2) carbons, and two methoxy groups (δ_C 56.8, 61.4; δ_H 3.41, 3.71). A direct comparison of ^{1}H and ^{13}C NMR spectra of **5** (Table 3, Figures S23 and S24) with those of terphenyllin [1] showed their close resemblance, with the exception of the presence of a hydroxy group at C-3$''$ (δ_C 146.4) and the difference in carbon chemical shifts at C-2$''$ (δ_C 117.5), C-4$''$ (δ_C 146.3), C-5$''$ (δ_C 116.8) and C-6$''$ (δ_C 122.0) (131.1 ppm for C-1$''$, C-2$''$, and C-6$''$; 158.0 ppm for C-4$''$ in terphenyllin [1]). The HMBC correlations (Figure 6, Figure S27) from H-2 (δ_H 7.25) to C-4 (δ_C 157.6), C-6 (δ_C 133.7), and C-4$'$ (δ_C 118.3); from H-3 (δ_H 6.85) to C-1 (δ_C 126.8) and C-5 (δ_C 115.9); from H-5 (δ_H 6.85) to C-3 (δ_C 115.9) and C-1; from H-6 (δ_H 7.25) to C-2 (δ_C 133.7), C-4, and C-4$'$; from H-6$'$ (δ_H 6.47) to C-2$'$ (δ_C 140.8), C-4$'$, and C-1$''$ (δ_C 131.9); from H-2$''$ (δ_H 7.19) to C-1$'$ (δ_C 134.2), C-4$''$, and C-6$''$; from H-5$''$ (δ_H 6.91) to C-1$''$ and C-3$''$; from H-6$''$ (δ_H 7.50) to C-1$'$, C-2$''$, and C-4$''$; and ROESY correlation from H-6$'$ and 2$'$-OMe (δ_H 3.41) to H-2$''$ and H-6$''$, and from 5$'$-OMe (δ_H 3.71) to H-2 (H-6) established the structure of **5** as the 2$'$,5$'$-dimethoxy-4,3$'$,3$''$,4$''$-tetrahydroxy-*p*-terphenyl derivative. This structure was previously published as hydroxyterphenyllin in an unavailable source [26]. It should be noted that these authors reported the name "hydroxyterphenyllin" for an isomeric compound [27] now known as 3-hydroxyterphenyllin [2,28]. Probably, the structure of **5** was mistakenly provided by [26] instead of the structure of 3-hydroxyterphenyllin. Therefore, **5** should be named 3$''$-hydroxyterphenyllin.

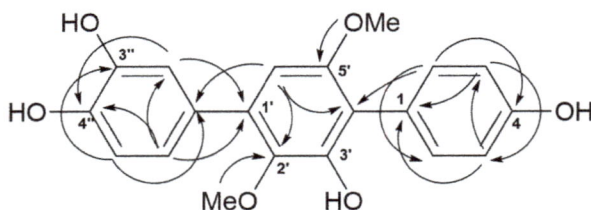

Figure 6. Key HMBC correlations of **5**.

Table 3. ^{1}H and ^{13}C NMR data (δ in ppm, DMSO-d_6) for 3″-hydroxyterphenyllin (**5**).

Position	δ_C, mult	δ_H (*J* in Hz)	HMBC	ROESY
1	126.8, C			
2	133.7, CH	7.25, d (8.4)	4, 6, 4′	
3	115.9, CH	6.85, d (8.5)	1, 5	
4	157.6, C			
5	115.9, CH	6.85, d (8.5)	1, 3	
6	133.7, CH	7.25, d (8.4)	2, 4, 4′	5′-OMe
1′	134.2, C			
2′	140.8, C			
3′	149.8, C			
4′	118.3, C			
5′	155.1, C			
6′	104.8, CH	6.47, s	2′, 4′, 1″	2″, 6″
1″	131.9, C			
2″	117.5, CH	7.19, d (2.1)	1′, 4″, 6″	6′, 2′-OMe
3″	146.4, C			
4″	146.3, C			
5″	116.8, CH	6.91, d (8.1)	1″, 3″	
6″	122.0, CH	7.02, dd (8.1, 2.1)	1′, 2″, 4″	6′, 2′-OMe
2′-OMe	61.4, CH$_3$	3.41, s	2′	2″, 6″
5′-OMe	56.8, CH$_3$	3.71, s	5′	6

^{1}H NMR and ^{13}C NMR spectroscopic data were measured at 700 MHz and 175 MHz, respectively.

The biosynthesis of related indole-diterpenes was previously proposed for paspalinine [29]. Apparently, the common biosynthetic precursor of asperindoles and 1,3-dioxolane indole-diterpenoids (including paspalinine) is 7α-hydroxypaxilline (Figure 7). Oxidation of the isopropyl substituent, followed by cyclization at C-7 and C-2′, generates a 1,3-dioxane ring. Asperindoles are then formed by acylation and halogenation.

The effect of the asperindoles A (**1**) and C (**3**) on cell viability, cell cycle progression, and induction of apoptosis in human prostate cancer cell lines was investigated. MTT assays revealed that asperindole C (**3**) was noncytotoxic against human PC-3, LNCaP (androgen-sensitive human prostate adenocarcinoma cells), and 22Rv1 cell lines with an IC$_{50}$ > 100 μM. In contrast, asperindole A (**1**) showed cytotoxicity in all three cell lines, with IC$_{50}$ values of 69.4 μM, 47.8 μM, and 4.86 μM, respectively. Docetaxel, which was used as a reference substance, displayed IC$_{50}$ values of 15.4 nM, 3.8 nM, and 12.7 nM, respectively. Asperindole A (**1**) was able to induce apoptosis in human cancer 22Rv1 cells at low-micromolar concentrations (Figure 8). Cell cycle progression analysis of 22Rv1 cells treated with asperindole A (**1**) for 48 h revealed a S-phase arrest (as well as a discrete G2/M-phase arrest, Figure 8). Thus, asperindole A (**1**) may be a promising candidate for further studies in human drug-resistant prostate cancer. In contrast, 22Rv1 cells treated with 100 μM of asperindole C (**3**) for 48 h revealed only minimal induction of apoptosis (8.9 ± 0.6% vs 1.2 ± 0.1% in the control) and no significant changes in cell cycle progression.

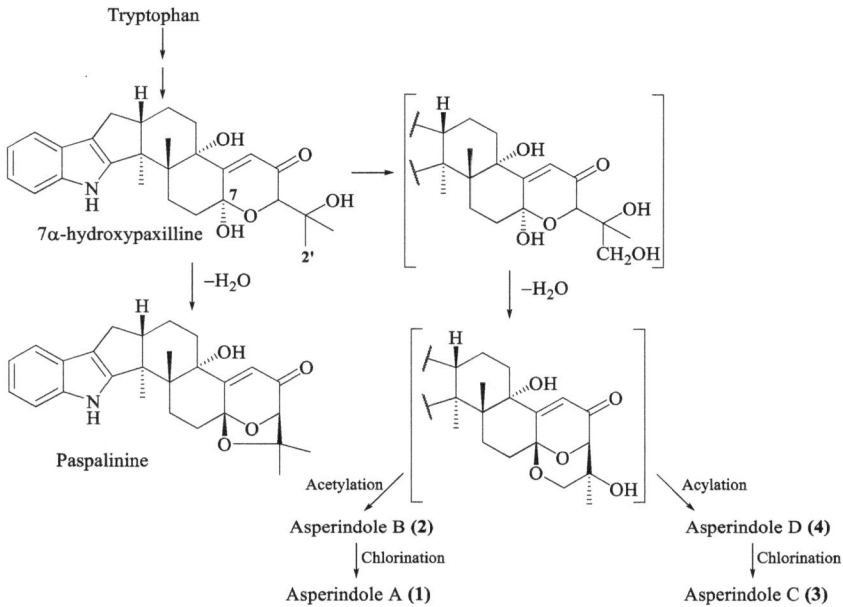

Figure 7. Proposed biosynthesis of asperindoles A–D (**1**–**4**).

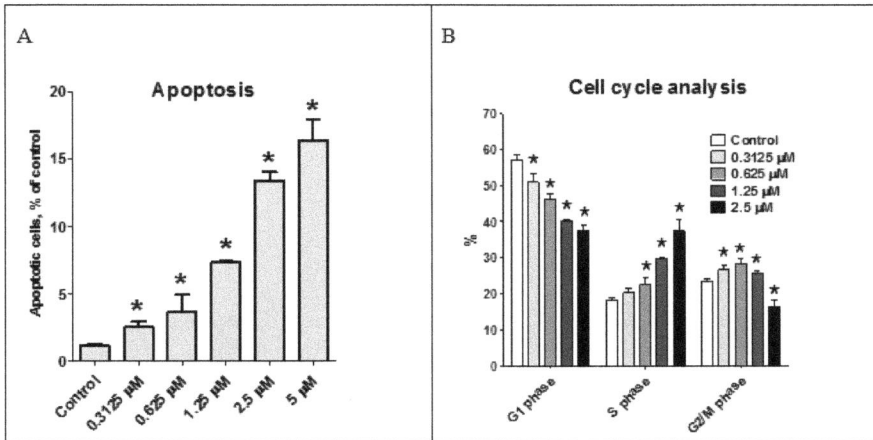

Figure 8. Effect of asperindole A (**1**) on cell cycle progression and apoptosis induction. Apoptotic cells were detected as a sub-G1 population (**A**); Cell cycle analysis of 22Rv1 cells treated with asperindole A (**1**) for 48 h (**B**). Cell cycle phase distribution, quantified using the Cell Quest Pro software. * $p < 0.05$.

3. Materials and Methods

3.1. General Experimental Procedures

Optical rotations were measured on a Perkin-Elmer 343 polarimeter (Perkin Elmer, Waltham, MA, USA). UV spectra were recorded on a Specord UV−vis spectrometer (Carl Zeiss, Jena, Germany) in CHCl$_3$. NMR spectra were recorded in DMSO-d_6 on a Bruker DPX-500 (Bruker BioSpin GmbH, Rheinstetten, Germany) and a Bruker DRX-700 (Bruker BioSpin GmbH, Rheinstetten, Germany) spectrometer, using TMS as an internal standard. HRESIMS spectra were measured on an Agilent 6510 Q-TOF LC mass spectrometer (Agilent Technologies, Santa Clara, CA, USA) and a Maxis impact mass spectrometer (Bruker Daltonics GmbH, Rheinstetten, Germany).

Low-pressure liquid column chromatography was performed using silica gel (50/100 µm, Imid, Russia). Plates (4.5 cm × 6.0 cm) precoated with silica gel (5–17 µm, Imid) were used for thin-layer chromatography. Preparative HPLC was carried out on a Shimadzu LC-20 chromatograph (Shimadzu USA Manufacturing, Canby, OR, USA) using a YMC ODS-AM (YMC Co., Ishikawa, Japan) (5 µm, 10 mm × 250 mm) and YMC SIL (YMC Co., Ishikawa, Japan) (5 µm, 10 mm × 250 mm) columns with a Shimadzu RID-20A refractometer (Shimadzu Corporation, Kyoto, Japan).

3.2. Fungal Strain

The strain was isolated from an unidentified colonial ascidian (Shikotan Island, Pacific Ocean) on malt extract agar, and identified on the basis of morphological and molecular features. For DNA extraction, the culture was grown on malt extract agar under 25 °C for 7 days. DNA extraction was performed with the HiPurATM Plant DNA Isolation kit (CTAB Method) (HiMedia Laboratories Pvt. Ltd., Mumbai, India) according to the manufacturer's instructions. Fragments containing the ITS (internal transcribed spacer) regions were amplified using ITS1 and ITS4 primers. The newly obtained sequences were checked visually and compared to available sequences in the GenBank database (www.mycobank.org). According to BLAST analysis of the ITS1–5.8S–ITS2 sequence, the strain KMM 4676 had 98% similarity with *Aspergillus candidus*. The sequences were deposited in the GenBank nucleotide sequence database under MG 241226. The strain is deposited in the Collection of Marine Microorganisms of G. B. Elyakov Pacific Institute of Bioorganic Chemistry FEB RAS under the code KMM 4676.

3.3. Cultivation of Fungus

The fungus was cultured at 22 °C for three weeks in 14 × 500 mL Erlenmeyer flasks, each containing rice (20.0 g), yeast extract (20.0 mg), KH$_2$PO$_4$ (10 mg), and natural sea water (40 mL).

3.4. Extraction and Isolation

The fungal mycelia with the medium were extracted for 24 h with 5.6 L of EtOAc. Evaporation of the solvent under reduced pressure gave a dark brown oil (6.25 g), to which 250 mL H$_2$O–EtOH (4:1) was added, and the mixture was thoroughly stirred to yield a suspension. It was extracted successively with *n*-hexane (150 mL × 2), EtOAc (150 mL × 2), and *n*-BuOH (150 mL × 2). After evaporation of the EtOAc layer, the residual material (3.92 g) was passed over a silica gel column (35.0 cm × 2.5 cm, 75 g), which was eluted first with *n*-hexane (1.0 L), followed by a step gradient from 5% to 100% EtOAc in *n*-hexane (total volume 30 L). Fractions of 250 mL each were collected and combined on the basis of TLC (Si gel, toluene–2-propanol, 6:1 and 3:1, *v/v*).

The *n*-hexane–EtOAc (9:1, 2 L, 21.70 mg) fraction was purified by LH-20 column (80 cm × 2 cm, 50 g) with CHCl$_3$ to yield 30 subfractions. Subfraction 6 is the individual compound **1** (8.30 mg). Subfractions 8–12 (7.00 mg) were purified by HPLC on a YMC ODS-AM column, eluting with MeOH–H$_2$O (9:1), and then by HPLC on a YMC SIL column, eluting with acetone–*n*-hexane (1:3) to yield **2** (0.56 mg), **3** (1.05 mg), and **4** (1.47 mg). The *n*-hexane–EtOAc (4:1, 2 L, 145 mg) fraction was purified by by HPLC on a YMC ODS-AM column, eluting with MeOH–H$_2$O (13:7), then by HPLC on a YMC-SIL column eluting first with CHCl$_3$–MeOH–NH$_4$OAc (90:10:1.5), and then with CHCl$_3$–MeOH–NH$_4$OAc (85:15:1) to yield **5** (56.10 mg).

Asperindole A (**1**): white powder; $[\alpha]_D^{20}$ +22 (*c* 0.10, CHCl$_3$); UV (MeOH) λ_{max} (log ε) 284 (3.86), 236.4 (4.55), 195.6 (4.55) nm; ECD (0.21 mM, MeOH) λ_{max} ($\Delta\varepsilon$) 205 (+8.52), 240 (−19.80), 280 (+0.25),

360 (+3.60) nm; ^{1}H and ^{13}C NMR data see Table 1, Figures S1–S6; HR ESIMS *m/z* 526.1980 [M + H]$^+$ (calcd. for C$_{29}$H$_{33}$NO$_6$Cl, 526.1992, Δ −2.28 ppm).

Asperindole B (**2**): white powder; $[\alpha]^{20}_D$ +40 (*c* 0.03, CHCl$_3$); ^{1}H and ^{13}C NMR data see Table 1, Figures S7–S13; HRESIMS *m/z* 514.2194 [M + Na]$^+$ (calcd. for C$_{29}$H$_{33}$NO$_6$Na, 514.2200, Δ −1.17 ppm).

Asperindole C (**3**): white powder; $[\alpha]^{20}_D$ +46 (*c* 0.72, CHCl$_3$); UV (MeOH) λ_{max} (log ε) 284 (3.90), 236.4 (4.56), 194.8 (4.46) nm; ECD (0.21 mM, MeOH) λ_{max} ($\Delta\varepsilon$) 205 (+6.43), 240 (−18.25), 280 (+0.02), 360 (+3.57) nm; ^{1}H and ^{13}C NMR data see Table 2, Figures S14–S20; HRESIMS *m/z* 610.2206 [M − H]$^-$ (calcd. for C$_{33}$H$_{37}$NO$_8$Cl, 610.2213, Δ −1.15 ppm).

Asperindole D (**4**): white powder; $[\alpha]^{20}_D$ +24 (*c* 0.05, CHCl$_3$); ^{1}H and ^{13}C NMR data, see Table 2, Figures S21 and S22; HRESIMS *m/z* 576.2594 [M − H]$^-$ (calcd. for C$_{33}$H$_{38}$NO$_8$, 576.2603, Δ −1.56 ppm).

3″-hydroxyterphenylline (**5**): ^{1}H and ^{13}C NMR data, see Table 3, Figures S23–S27; HRESIMS *m/z* 353.1036 [M − H]$^-$ (calcd. for C$_{20}$H$_{17}$O$_6$, 353.1031, Δ −1.42 ppm).

3.5. Cell Culture

The human prostate cancer cells lines 22Rv1, PC-3, and LNCaP were purchased from ATCC. Cell lines were cultured in 10% FBS/RPMI media (Invitrogen Ltd., Paisley, UK) with (for LNCaP) or without (for 22Rv1 and PC-3) 1 mM sodium pyruvate (Invitrogen). Cells were continuously kept in culture for a maximum of 3 months, and were routinely inspected microscopically for stable phenotype and regularly checked for contamination with mycoplasma. Cell line authentication was performed by DSMZ (Braunschweig, Germany) using highly polymorphic short tandem repeat loci [30].

3.6. Cytotoxicity Assay

The in vitro cytotoxicity of individual substances was evaluated using the MTT (3-(4,5-dimethylthiazol-2-yl)-2,5-diphenyltetrazolium bromide) assay, which was performed as previously described [31]. Docetaxel was used as a control.

3.7. Cell Cycle and Apoptosis Induction Analysis

The cell cycle distribution was analyzed by flow cytometry using PI (propidium iodide) staining as described before with slight modifications [32]. In brief, cells were preincubated overnight in 6-well plates (2×10^5 cells/well in 2 mL/well). The medium was changed to fresh medium containing different concentrations of the substances. After 48 h of treatment, cells were harvested with a trypsin-EDTA solution, fixed with 70% EtOH, stained, and analyzed by BD Bioscience FACS Calibur analyzer (BD Bioscience, Bedford, MA, USA). The results were quantitatively analyzed using BD Bioscience Cell Quest Pro v.5.2.1. software (San Jose, CA, USA). Cells detected in the sub-G1 peak were considered as apoptotic.

4. Conclusions

Four new metabolites, the indole-diterpene alkaloids asperindoles A–D (**1–4**), and the known *p*-terphenyl derivative 3″-hydroxyterphenyllin (**5**) were isolated from a marine-derived strain of the fungus *A. candidus* KMM 4676, associated with an unidentified colonial ascidian. To the best of our knowledge, **3** and **4** are the first examples of naturally occurring compounds containing a 2-hydroxyisobutiric acid (2-HIBA) residue. This is the first report of the spectral data and reliable assignment for 3″-hydoxyterphenyllin (**5**). Asperindole A (**1**) was proved to be highly cytotoxic in 22Rv1 human prostate cancer cells resistant to androgen receptor-targeted therapies. Therefore, this compound is a promising candidate for further evaluation in human drug-resistant prostate cancer cells.

Supplementary Materials: ^{1}H, ^{13}C, DEPT, COSY-45, HSQC, HMBC, and ROESY spectra of new compounds **1–5** are available online at http://www.mdpi.com/1660-3397/16/7/232/s1.

Author Contributions: Conceptualization, A.N.Y.; Data curation, A.N.Y. and S.A.D.; Formal analysis, A.N.Y.; Funding acquisition, G.v.A. and S.S.A.; Investigation, E.V.I., A.N.Y., O.F.S., A.B.R., O.I.Z., M.V.P., R.S.P. and S.A.D.; Methodology, A.N.Y., G.v.a. and S.A.D.; Project administration, G.v.A. and S.S.A.; Resources, A.N.Y. and G.v.a.; Supervision, S.S.A.; Validation, S.A.D.; Visualization, E.V.I.; Writing-original draft, E.V.I., M.V.P. and S.A.D.; Writing-review & editing, A.N.Y., G.v.a., S.S.A. and S.A.D.

Acknowledgments: The study was supported by the Federal Agency for Scientific Organizations program for support to bioresource collections. The authors thank Friedemann Honecker for thorough English editing.

Conflicts of Interest: The authors declare no conflict of interest.

References

1. Marchelli, R.; Vining, L.C. Terphenyllin, a novel p terphenyl metabolite from *Aspergillus candidus*. *J. Antibiot.* **1975**, *28*, 328–331. [CrossRef] [PubMed]

2. Kurobane, I.; Vining, L.C.; McInnes, A.G.; Smith, D.G. 3-Hydroxyterphenyllin, a new metabolite of *Aspergillus candidus*. Structure elucidation by 1H and 13C nuclear magnetic resonance spectroscopy. *J. Antibiot.* **1979**, *32*, 559–564. [CrossRef] [PubMed]

3. Kamigauchi, T.; Sakazaki, R.; Nagashima, K.; Kawamura, Y.; Yasuda, Y.; Matsushima, K.; Tani, H.; Takahashi, Y.; Ishii, K.; Suzuki, R.; et al. Terprenins, novel immunosuppressants produced by *Aspergillus candidus*. *J. Antibiot.* **1998**, *51*, 445–450. [CrossRef] [PubMed]

4. Rahbæk, L.; Frisvad, J.C.; Christophersen, C. An amendment of *Aspergillus* section *Candidi* based on chemotaxonomical evidence. *Phytochemistry* **2000**, *53*, 581–586. [CrossRef]

5. Cai, S.; Sun, S.; Zhou, H.; Kong, X.; Zhu, T.; Li, D.; Gu, Q. Prenylated polyhydroxy-p-terphenyls from *Aspergillus taichungensis* ZHN-7-07. *J. Nat. Prod.* **2011**, *74*, 1106–1110. [CrossRef] [PubMed]

6. Bird, A.E.; Marshall, A.C. Structure of chlorflavonin. *J. Chem. Soc. C* **1969**, *18*, 2418–2420. [CrossRef]

7. Yan, T.; Guo, Z.K.; Jiang, R.; Wei, W.; Wang, T.; Guo, Y.; Song, Y.C.; Jiao, R.H.; Tan, R.X.; Ge, H.M. New Flavonol and Diterpenoids from the Endophytic Fungus *Aspergillus* sp. YXf3. *Planta Med.* **2013**, *79*, 348–352. [CrossRef] [PubMed]

8. Kuhnert, E.; Surup, F.; Herrmann, J.; Huch, V.; Müller, R.; Stadler, M.; Rickenyls, A.-E. antioxidative terphenyls from the fungus Hypoxylon rickii (Xylariaceae, Ascomycota). *Phytochemistry* **2015**, *118*, 68–73. [CrossRef] [PubMed]

9. Valeria Calì, C.S.; Tringali, C. Polyhydroxy-P-Terphenyls and Related P-Terphenylquinones From Fungi: Overview and Biological Properties. *Stud. Nat. Prod. Chem.* **2003**, *29*, 263–307.

10. Kobayashi, A.; Takemoto, A.; Koshimizu, K.; Kawazu, K. p-Terphenyls with cytotoxic activity toward sea urchin embryos. *Agric. Biol. Chem.* **1985**, *49*, 867–868. [CrossRef]

11. Buttachon, S.; Ramos, A.A.; Inácio, Â.; Dethoup, T.; Gales, L.; Lee, M.; Costa, P.M.; Silva, A.M.S.; Sekeroglu, N.; Rocha, E.; et al. Bis-indolyl benzenoids, hydroxypyrrolidine derivatives and other constituents from cultures of the marine sponge-associated fungus *Aspergillus candidus* KUFA0062. *Mar. Drugs* **2018**, *16*, 119. [CrossRef] [PubMed]

12. Netz, N.; Opatz, T. Marine indole alkaloids. *Mar. Drugs* **2015**, *13*, 4814–4914. [CrossRef] [PubMed]

13. Munday-Finch, S.C.; Miles, C.O.; Wilkins, A.L.; Hawkes, A.D. Isolation and Structure Elucidation of Lolitrem A, a Tremorgenic Mycotoxin from *Perennial Ryegrass* Infected with *Acremonium lolii*. *J. Agric. Food Chem.* **1995**, *43*, 1283–1288. [CrossRef]

14. Gallagher, R.T.; White, E.P.; Mortimer, P.H. Ryegrass staggers: Isolation of potent neurotoxins lolitrem a and lolitrem b from staggers-producing pastures. *N. Z. Vet. J.* **1981**, *29*, 189–190. [CrossRef] [PubMed]

15. Munday-Finch, S.C.; Wilkins, A.L.; Miles, C.O.; Tomoda, H.; Omura, S. Isolation and Structure Elucidation of Lolilline, a Possible Biosynthetic Precursor of the Lolitrem Family of Tremorgenic Mycotoxins. *J. Agric. Food Chem.* **1997**, *45*, 199–204. [CrossRef]

16. Gallagher, R.T.; Finer, J.; Clardy, J.; Leutwiler, A.; Weibel, F.; Acklin, W.; Arigoni, D. Paspalinine, a Tremorgenic Metabolite from *Claviceps paspali* Stevens Et Hall. *Tetrahedron Lett.* **1980**, *21*, 235–238. [CrossRef]

17. Sun, K.; Li, Y.; Guo, L.; Wang, Y.; Liu, P.; Zhu, W. Indole diterpenoids and isocoumarin from the fungus, *Aspergillus flavus*, isolated from the prawn, *Penaeus vannamei*. *Mar. Drugs* **2014**, *12*, 3970–3981. [CrossRef] [PubMed]

18. Fan, Y.; Wang, Y.; Liu, P.; Fu, P.; Zhu, T.; Wang, W.; Zhu, W. Indole-diterpenoids with anti-H1N1 activity from the aciduric fungus *Penicillium camemberti* OUCMDZ-1492. *J. Nat. Prod.* **2013**, *76*, 1328–1336. [CrossRef] [PubMed]

19. Belofsky, G.N.; Gloer, J.B.; Wicklow, D.T.; Dowd, P.F. Antiinsectan alkaloids: Shearinines A-C and a new paxilline derivative from the ascostromata of *Eupenicillium shearii*. *Tetrahedron* **1995**, *51*, 3959–3968. [CrossRef]

20. Harms, H.; Rempel, V.; Kehraus, S.; Kaiser, M.; Hufendiek, P.; Müller, C.E.; König, G.M. Indoloditerpenes from a marine-derived fungal strain of *Dichotomomyces cejpii* with antagonistic activity at GPR18 and cannabinoid receptors. *J. Nat. Prod.* **2014**, *77*, 673–677. [CrossRef] [PubMed]

21. Yurchenko, A.N.; Ivanets, E.V.; Smetanina, O.F.; Pivkin, M.V.; Dyshlovoi, S.A.; von Amsberg, G.; Afiyatullov, S.S. Metabolites of the Marine Fungus *Aspergillus candidus* KMM 4676 Associated with a Kuril Colonial Ascidian. *Chem. Nat. Compd.* **2017**, *53*, 747–749. [CrossRef]

22. Sallam, A.A.; Houssen, W.E.; Gissendanner, C.R.; Orabi, K.Y.; Foudah, A.I.; El Sayed, K.A. Bioguided discovery and pharmacophore modeling of the mycotoxic indole diterpene alkaloids penitrems as breast cancer proliferation, migration, and invasion inhibitors. *MedChemComm* **2013**, *4*. [CrossRef] [PubMed]

23. De Jesus, A.E.; Steyn, P.S.; van Heerden, F.R.; Vleggaar, R.; Wessels, P.L.; Hull, W.E. Tremorgenic mycotoxins from *Penicillium crustosum*. Structure elucidation and absolute configuration of penitrems B–F. *J. Chem. Soc. Perkin Trans. 1* **1983**, *41*, 1857–1861. [CrossRef]

24. De Jesus, A.E.; Steyn, P.S.; Van Heerden, F.R.; Vleggaar, R.; Wessels, P.L.; Hull, W.E. Tremorgenic mycotoxins from *Penicillium crustosum*: Isolation of penitrems A–F and the structure elucidation and absolute configuration of penitrem A. *J. Chem. Soc. Perkin Trans. 1* **1983**, *14*, 1847–1856. [CrossRef]

25. Penn, J.; Biddle, J.R.; Mantle, P.G.; Bilton, J.N.; Sheppard, R.N. Pennigritrem, a naturally-occurring penitrem A analogue with novel cyclisation in the diterpenoid moiety. *J. Chem. Soc. Perkin Trans. 1* **1992**, *23*, 23–26. [CrossRef]

26. Cutler, H.G.; Cole, R.J.; Cox, R.H.; Wells, J.M. Fungal metabolites: Interesting new plant growth inhibitors. *Proc. Plant Growth Regul. Work. Group* **1979**, *6*, 87–91.

27. Cutler, H.G.; LeFiles, J.H.; Crumley, F.G.; Cox, R.H. Hydroxyterphenyllin: A novel fungal metabolite with plant growth inhibiting properties. *J. Agric. Food Chem.* **1978**, *26*, 632–635. [CrossRef]

28. Guo, Z.K.; Yan, T.; Guo, Y.; Song, Y.C.; Jiao, R.H.; Tan, R.X.; Ge, H.M. P-terphenyl and diterpenoid metabolites from endophytic *Aspergillus* sp. YXf3. *J. Nat. Prod.* **2012**, *75*, 15–21. [CrossRef] [PubMed]

29. Mantle, P.G.; Weedon, C.M. Biosynthesis and transformation of tremorgenic indole-diterpenoids by *Penicillium paxilli* and *Acremonium lolii*. *Phytochemistry* **1994**, *36*, 1209–1217. [CrossRef]

30. Dyshlovoy, S.A.; Menchinskaya, E.S.; Venz, S.; Rast, S.; Amann, K.; Hauschild, J.; Otte, K.; Kalinin, V.I.; Silchenko, A.S.; Avilov, S.A.; et al. The marine triterpene glycoside frondoside A exhibits activity in vitro and in vivo in prostate cancer. *Int. J. Cancer* **2016**, *138*, 2450–2465. [CrossRef] [PubMed]

31. Dyshlovoy, S.A.; Venz, S.; Shubina, L.K.; Fedorov, S.N.; Walther, R.; Jacobsen, C.; Stonik, V.A.; Bokemeyer, C.; Balabanov, S.; Honecker, F. Activity of aaptamine and two derivatives, demethyloxyaaptamine and isoaaptamine, in cisplatin-resistant germ cell cancer. *J. Proteom.* **2014**, *96*, 223–239. [CrossRef] [PubMed]

32. Dyshlovoy, S.A.; Hauschild, J.; Amann, K.; Tabakmakher, K.M.; Venz, S.; Walther, R.; Guzii, A.G.; Makarieva, T.N.; Shubina, L.K.; Fedorov, S.N.; et al. Marine alkaloid monanchocidin a overcomes drug resistance by induction of autophagy and lysosomal membrane permeabilization. *Oncotarget* **2015**, *6*, 17328–17341. [CrossRef] [PubMed]

marine drugs

MDPI

Communication

A New Breviane Spiroditerpenoid from the Marine-Derived Fungus *Penicillium* sp. TJ403-1

Beiye Yang [1,†], Weiguang Sun [1,†], Jianping Wang [1,†], Shuang Lin [1], Xiao-Nian Li [2], Hucheng Zhu [1], Zengwei Luo [1], Yongbo Xue [1], Zhengxi Hu [1,*] and Yonghui Zhang [1,*]

[1] Hubei Key Laboratory of Natural Medicinal Chemistry and Resource Evaluation, School of Pharmacy, Tongji Medical College, Huazhong University of Science and Technology, Wuhan 430030, China; yangbeiye123@163.com (B.Y.); weiguang_s@hust.edu.cn (W.S.); jpwang1001@163.com (J.W.); 13207186519@163.com (S.L.); zhuhucheng0@163.com (H.Z.); luozengwei@gmail.com (Z.L.); yongboxue@mail.hust.edu.cn (Y.X.)

[2] State Key Laboratory of Phytochemistry and Plant Resources in West China, Kunming Institute of Botany, Chinese Academy of Sciences, Kunming 650204, China; lixiaonian@mail.kib.ac.cn

* Correspondence: huzhengxi@hust.edu.cn (Z.H.); zhangyh@mails.tjmu.edu.cn (Y.Z.); Tel.: +86-027-83692892 (Y.Z.)

† These authors contributed equally to this work.

Received: 5 March 2018; Accepted: 28 March 2018; Published: 29 March 2018

Abstract: Marine-derived fungi are a promising and untapped reservoir for discovering structurally interesting and pharmacologically active natural products. In our efforts to identify novel bioactive compounds from marine-derived fungi, four breviane spiroditerpenoids, including a new compound, brevione O (**1**), and three known compounds breviones I (**2**), J (**3**), and H (**4**), together with a known diketopiperazine alkaloid brevicompanine G (**5**), were isolated and identified from an ethyl acetate extract of the fermented rice substrate of the coral-derived fungus *Penicillium* sp. TJ403-1. The absolute structure of **1** was elucidated by HRESIMS, one- and two-dimensional NMR spectroscopic data, and a comparison of its electronic circular dichroism (ECD) spectrum with the literature. Moreover, we confirmed the absolute configuration of **5** by single-crystal X-ray crystallography. All the isolated compounds were evaluated for isocitrate dehydrogenase 1 (IDH1) inhibitory activity and cytotoxicity, and compound **2** showed significant inhibitory activities against HL-60, A-549, and HEP3B tumor cell lines with IC_{50} values of 4.92 ± 0.65, 8.60 ± 1.36, and 5.50 ± 0.67 μM, respectively.

Keywords: marine-derived fungi; *Penicillium* sp. TJ403-1; breviane spiroditerpenoid; IDH1 inhibitory activity; cytotoxicity

1. Introduction

Since the ocean covers over 70% of the Earth's surface, marine organisms are regarded as a prolific and under-explored resource of bioactive natural products [1,2]. Over the past few decades, with the discovery of plenty of new chemicals, marine-derived fungi have increasingly attracted the attention of natural product chemists and biologists, largely due to their surprising potentials for drug discovery [3–5].

Breviane spiroditerpenoids, which are biosynthesized from geranylgeranyl and pyrone derived from three molecules of acetyl CoA and one methyl from methionine, are an important group of architecturally complex and bioactive meroterpenoids [6]. Since the first breviane spiroditerpenoid, namely brevione A, was found from *Penicillium brevicompactum* Dierckx in 2000 [7], a total of 14 naturally occurring compounds with identical skeletons, showing intriguing allelopathic, anti-HIV, cytotoxic, and Aβ aggregate-induced neurotoxic inhibitory effects have been reported to date [7–11].

Notably, their architecturally complex frameworks with multiple chiral centers and attractive biological profiles made these natural products target molecules for total synthesis [12–14].

Previously, we performed a chemical investigation on a mangrove-derived fungus, *Daldinia eschscholzii*, resulting in the isolation and identification of three novel polyketide glycosides, daldinisides A–C, and two new alkaloids, 1-(3-indolyl)-2R,3-dihydroxypropan-1-one and 3-ethyl-2,5-pyrazinedipropanoic acid [15]. Interestingly, daldinisides A–C represented a rare class of D-ribose-containing natural products. In our continuous exploration for structurally unique and biologically active natural products from the marine-derived fungi, our attention was focused on the coral-derived fungi, a neglected and insufficiently explored natural resource. A chemical investigation of the fermented rice substrate of the coral-derived fungus *Penicillium* sp. TJ403-1 afforded four breviane spiroditerpenoids, including a new compound, brevione O (**1**), and three known compounds breviones I (**2**), J (**3**), and H (**4**), together with a known diketopiperazine alkaloid brevicompanine G (**5**). Herein, the details of the isolation, structural elucidation, and bioactivity evaluations of these compounds (Figure 1) are reported.

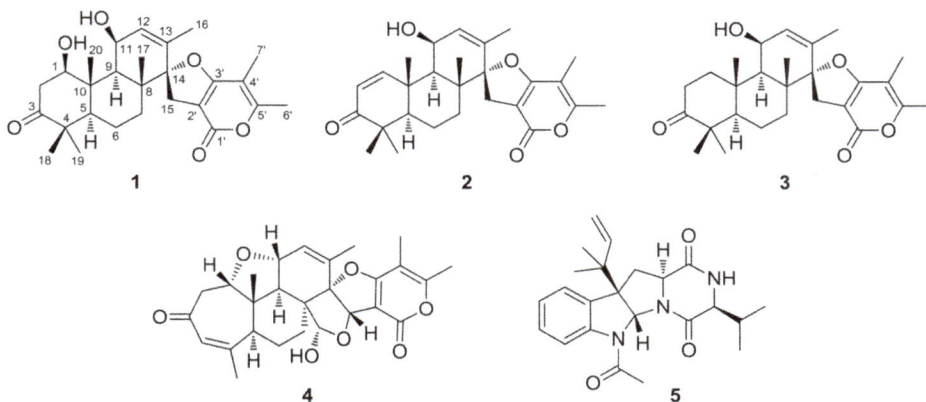

Figure 1. Structures of compounds **1–5**.

2. Results

Brevione O (**1**), purified as a white powder, was assigned the molecular formula $C_{27}H_{36}O_6$, based on the HRESIMS m/z 479.2382 [M + Na]$^+$ (calcd. for $C_{27}H_{36}O_6Na$, 479.2410) and ^{13}C-NMR analysis (see Supplementary Materials), corresponding to 10 degrees of unsaturation. Its ^1H-NMR data (Table 1) showed signals of seven methyl groups (δ_H 1.76 (d, J = 1.6 Hz, H$_3$-16), 1.21 (s, H$_3$-17), 1.09 (s, H$_3$-18), 1.03 (s, H$_3$-19), 1.48 (s, H$_3$-20), 1.92 (s, H$_3$-6′), and 2.25 (s, H$_3$-7′)), one olefinic proton (δ_H 5.74 (dd, J = 1.6, 4.8 Hz, H-12)), and two oxygenated methine protons (δ_H 3.95 (dd, J = 6.2, 9.2 Hz, H-1) and 4.68 (m, H-11)). Detailed analysis of the ^{13}C- and DEPT NMR spectroscopic data (Table 1) of **1** indicated the presence of seven sp^3 methyls, four sp^3 methylenes, five methines (including two oxygenated and one olefinic), eleven non-protonated carbons (including one oxygenated, five olefinic, one ester carbonyl, and one ketone). Comparing the HRESIMS and NMR data with previously reported C_{27} meroterpenoids from the *Penicillium* species [7–11], these data were indicative of a breviane spiroditerpenoid, and interpretations of the ^1H–^1H COSY and HMBC spectra (Figure 2) of **1** established its planar structure as the C-1 hydroxylated analogue of the known compound brevione J (**3**), a co-isolated known metabolite that was initially characterized from a marine-derived *Penicillium* sp. [10].

In the NOESY experiment (Figure 2), the cross-peaks of H$_3$-18/H-2β (δ_H 2.81), H-2β/H$_3$-20, H$_3$-20/H$_3$-17, and H$_3$-19/H-5, H-5/H-9, H-9/H-11, and correlations of H-1 with H-5, H-9, and H-11 indicated that H_3-17, H_3-18, and H_3-20 were all β-oriented, while H-1, H-5, H-9, H-11, and H_3-19 were all on the opposite side with α-orientations. Moreover, the NOE correlations of H_2-15 with H-7β (δ_H 1.62), H_3-16, and H_3-17 indicated that two planes of C- and D-rings were vertically arranged and that C-15 was on the upside of C-ring. Thus, the relative configuration of compound **1** was established.

To determine its absolute stereochemistry, the experiment electronic circular dichroism (ECD) spectrum (Figure 3) of **1**, which was measured in MeOH, was compared to that of brevione J (**3**). Their nearly identical ECD spectra indicated that the absolute stereochemistry of **1** should be identified as 1*R*,5*R*,8*R*,9*R*,10*S*,11*S*,14*S*.

By comparing their specific rotation and NMR data with the literature, compounds **2–5** were identified as breviones I (**2**) [10], J (**3**) [10], H (**4**) [9], and brevicompanine G (**5**) [16], respectively. Remarkably, the absolute configuration of **5** was first confirmed via single-crystal X-ray crystallography with a Flack parameter = −0.03(4) (Figure 4).

Table 1. ^{1}H- and ^{13}C-NMR data for brevione O (**1**) in methanol-d_4 (δ in ppm, *J* in Hz).

No.	1	
	δ_H [a,b]	δ_C [c]
1	3.95 dd (6.2, 9.2)	78.4 CH
2	2.64 dd (6.2, 14.6); 2.81 dd (9.2, 14.6)	45.0 CH_2
3	-	216.2 C
4	-	48.5 C
5	1.20 m	53.8 CH
6	1.66 m; 1.91 m	20.1 CH_2
7	1.39 m; 1.62 m	32.5 CH_2
8	-	41.7 C
9	1.78 d (5.7)	52.6 CH
10	-	44.8 C
11	4.68 m	67.6 CH
12	5.74 dd (1.6, 4.8)	131.1 CH
13	-	134.0 C
14	-	101.1 C
15	3.05 s	30.1 CH_2
16	1.76 d (1.6)	19.0 CH_3
17	1.21 s	19.1 CH_3
18	1.09 s	21.7 CH_3
19	1.03 s	26.1 CH_3
20	1.48 s	14.4 CH_3
1′	-	173.5 C
2′	-	100.6 C
3′	-	164.2 C
4′	-	104.8 C
5′	-	162.2 C
6′	1.92 s	9.5 CH_3
7′	2.25 s	17.2 CH_3

[a] Recorded at 400 MHz; [b] "m" means overlapped or multiplet with other signals; [c] Recorded at 100 MHz.

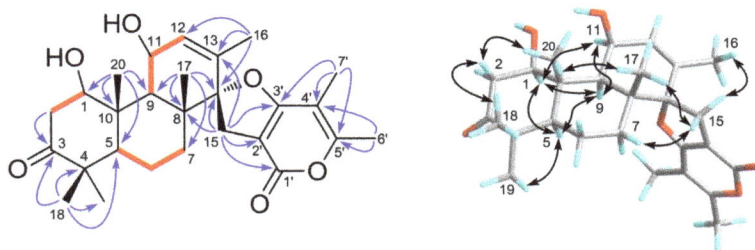

Figure 2. Selected ^{1}H–^{1}H COSY (red lines), HMBC (blue arrows), and NOESY (black arrows) correlations of compound **1**.

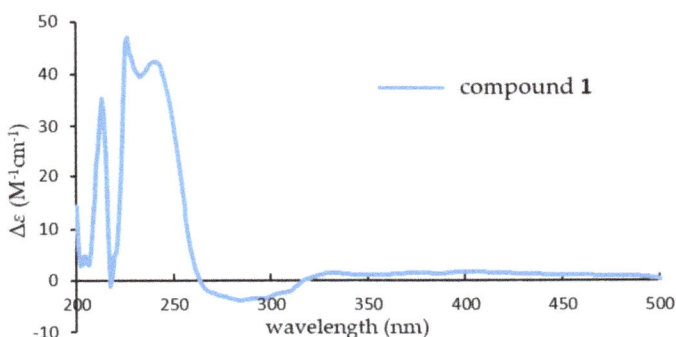

Figure 3. Experimental electronic circular dichroism (ECD) spectrum of compound **1**.

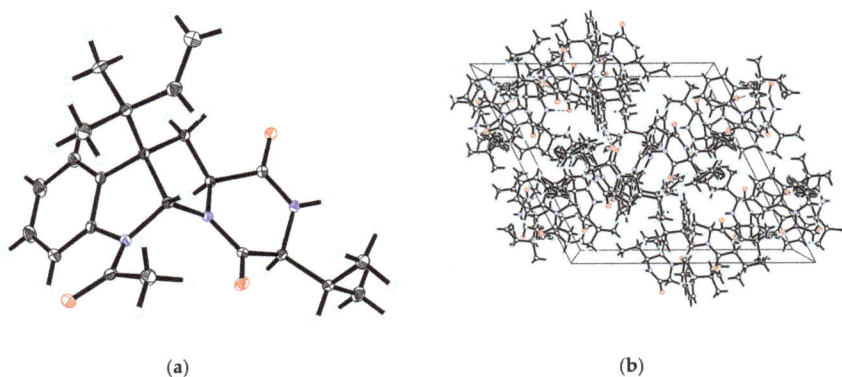

(a)   (b)

Figure 4. (**a**) ORTEP drawing of compound **5**; (**b**) View of the pack drawing of **5** and hydrogen-bonds are shown as dashed lines.

Isocitrate dehydrogenase 1 (IDH1) catalyzes the oxidative decarboxylation of isocitrate to α-ketoglutarate (α-KG). Neomorphic mutations in IDH1, mostly occurring at arginine 132, are frequently found in several human cancer types, including glioma, acute myeloid leukemia (AML), and myeloproliferative neoplasm [17]. When our research group screened for the IDH1 inhibitors from our natural product libraries, all the isolates **1**–**5** were evaluated for IDH1 inhibitory activity;

unfortunately, none of them was active at a concentration of 20 μM. Additionally, compounds **1–5** were investigated for cytotoxic activities against several human tumor cell lines, including HL-60 (acute leukemia), MM231 (breast cancer), A-549 (lung cancer), HEP3B (hepatic cancer), SW-480 (colon cancer), and one normal colonic epithelial cell, NCM460. As shown in Table 2, compound **2**, with no cytotoxic activity against the NCM460 cell, showed significant inhibitory potency against the HL-60, A-549, and HEP3B cell lines, with IC_{50} values of 4.92 ± 0.65, 8.60 ± 1.36, and 5.50 ± 0.67 μM, respectively. However, other compounds showed no obvious cytotoxicity against all these cell lines. Comparison of the IC_{50} values of **2** and **3** implied that the α,β-unsaturated carbonyl group plays an essential role in the cytotoxic activity. This group might interact with biological molecules by forming covalent bonds with free thiol of cysteine [18]; thus, the potential mechanism deserves further investigation in a following study.

Table 2. Cytotoxic activities of compounds **1–5** against human tumor cell lines.

Compound	IC_{50} (μM)					
	HL-60	MM231	A-549	HEP3B	SW480	NCM460
1	>40	>40	>40	>40	>40	>40
2	4.92 ± 0.65	>40	8.60 ± 1.36	5.50 ± 0.67	21.17 ± 2.72	>40
3	25.63 ± 3.84	>40	>40	>40	>40	>40
4	>40	>40	>40	>40	>40	>40
5	21.77 ± 3.48	>40	18.41 ± 2.15	>40	>40	>40
Cis-platin [1]	1.63 ± 0.11	3.84 ± 0.25	2.79 ± 0.36	2.96 ± 0.22	1.42 ± 0.11	0.85 ± 0.07
Paclitaxel [1]	<0.008	<0.008	<0.008	<0.008	<0.008	<0.008

[1] *Cis*-platin and paclitaxel were used as positive controls.

3. Materials and Methods

3.1. General Experimental Procedures

Optical rotations, ultraviolet (UV), and ECD data were recorded on a PerkinElmer 341 polarimeter (Waltham, MA, USA), a Varian Cary 50 (Santa Clara, CA, USA), and a JASCO-810 CD (Oklahoma City, OK, USA) spectrometer instrument, respectively. FT-IR spectra were measured by using a Bruker Vertex 70 instrument (Billerica, MA, USA). One- and two-dimensional NMR data were collected from a Bruker AM-400 instrument (Bruker, Karlsruhe, Germany). The ^{1}H- and ^{13}C-NMR chemical shifts of the solvent peaks for methanol-d_4 (δ_H 3.31 and δ_C 49.0) were referenced. The positive ion mode on a Thermo Fisher LC-LTQ-Orbitrap XL instrument (Waltham, MA, USA) was used for the measurement of high-resolution electrospray ionization mass spectra (HRESIMS). Semi-preparative HPLC separations were performed on an Agilent 1200 instrument with a reversed-phased C_{18} column (5 μm, 10 × 250 mm), using a UV detector or a Dionex HPLC system (Sunnyvale, CA, USA) which was equipped with an Ultimate 3000 autosampler injector, an Ultimate 3000 pump, and an Ultimate 3000 DAD controlled by Chromeleon software (version 6.80). Column chromatography (CC) was performed using silica gel (200−300 mesh, Qingdao Marine Chemical, Inc., Qingdao, China), Sephadex LH-20 (GE Healthcare Bio-Sciences AB, Uppsala, Sweden), and Lichroprep RP-C_{18} gel (40–63 μm, Merck, Darmstadt, Germany). Thin-layer chromatography was performed with the silica gel 60 F_{254} and RP-C_{18} F_{254} plates (Merck, Darmstadt, Germany), and spots were visualized by spraying heated silica gel plates with 10% H_2SO_4 in EtOH.

3.2. Fungal Material

Strain TJ403-1 was isolated from a piece of the inner tissues of a fresh soft coral of the genus *Alcyonium* (an unidentified sp.), which was collected from the Sanya Bay, Hainan Island, China. According to its morphology and sequence analysis of the ITS (Internal Transcribed Spacer) region of the rDNA, the strain was identified as *Penicillium* sp. and its sequence data have been submitted to the

GenBank with the accession no. MG839539. This strain has been reserved in the culture collection of Tongji Medical College, Huazhong University of Science and Technology (Wuhan, China).

3.3. Cultivation, Extraction, and Isolation

The TJ403-1 strain was incubated on potato dextrose agar (PDA) at 28 °C for 6 days in a stationary phase to prepare the seed cultures, which were then cut into small pieces (about $0.4 \times 0.4 \times 0.4$ cm) and inoculated into 350×500 mL sterilized Erlenmeyer flasks (each comprised of 200 g rice and 200 mL distilled water). After an incubation at 28 °C for 40 days, 300 mL EtOAc was added to all flasks to stop the growth of cells. Then, the filtrates were collected, and the fermented rice substrate was extracted with EtOAc (8×30 L) six times. Lastly, these two parts of the extracts were mixed together. Under reduced pressure, the organic solvent was evaporated to dryness to obtain a dark brown crude extract (700 g).

The EtOAc extract (700 g) was subjected to silica gel CC eluted with a gradient of petroleum ether/EtOAc/MeOH (stepwise 20:1:0 to 1:1:1, $v/v/v$) to afford eight fractions (Fr.1–Fr.8). Fr.5 was separated through RP-C_{18} CC (MeOH/H_2O, 20% to 100%, v/v) to give five subfractions (Fr.5.1–Fr.5.5). Then, Fr.5.3 was purified via silica gel CC, eluted with CH_2Cl_2/MeOH (stepwise 50:1 to 30:1, v/v) to afford two additional subfractions (Fr.5.3.1−5.3.2). Further separations of Fr.5.3.1 through Sephadex LH-20 eluted with a mixture of CH_2Cl_2/MeOH (1:1, v/v) and repeated semi-preparative HPLC (MeOH/H_2O, 70:30, v/v; 2 mL/min) yielded compounds **2** (16 mg), **3** (8 mg), and **5** (9 mg). Fr.6 was loaded onto RP-C_{18} CC (MeOH/H_2O, 20% to 100%, v/v) to give five subfractions (Fr.6.1–Fr.6.5). By using the silica gel CC (CH_2Cl_2/MeOH, 40:1, v/v) and repeated semi-preparative HPLC (MeOH/H_2O, 60:40, v/v; 2 mL/min) methods, Fr.6.3 was finally separated to yield compounds **1** (5 mg) and **4** (60 mg).

Brevione O (**1**): white powder; $[\alpha]_D^{23}$: +94.1 (*c* 0.14, MeOH); UV (MeOH) λ_{max} (log ε): 218 (4.33), 253 (3.73), 307 (3.46) nm; CD (MeOH) λ_{max} ($\Delta\varepsilon$): 213 (+35.24), 225 (+45.98), 238 (+41.91), 283 (−3.94); IR (KBr) ν_{max}: 3423, 2925, 2855, 1703, 1573, 1446, 1388, 1363, 1275, 1116, 1062, 983, 924, 856, 746 cm^{-1}; HRESIMS *m*/*z* 479.2382 [M + Na]$^+$ (calcd. for $C_{27}H_{36}O_6Na$, 479.2410). For ^{1}H- and ^{13}C-NMR data, see Table 1.

3.4. X-ray Crystallographic Analysis

At room temperature, after attempts with various organic solvents, compound **5** was obtained as colorless crystals from MeOH/H_2O (15:1, v/v) by slow evaporation. The intensity data for compound **5** was collected with a Bruker APEX DUO diffractometer, which was outfitted with an APEX II CCD by using graphite-monochromated Cu Kα radiation (100 K). The Bruker SAINT was used for the cell refinement and data reduction of compound **5**, whose structure was then solved and refined through direct means by using SHELXS-97 program (Göttingen, Germany) [19]. The crystallographic data for compound **5** have been deposited in the Cambridge Crystallographic Data Center (CCDC 1827348 for **5**). Copies of the data can be obtained free of charge from the CCDC, 12 Union Road, Cambridge CB 1EZ, UK (fax: Int. +44(0) (1223) 336 033); e-mail: deposit@ccdc.cam.ac.uk).

Crystallographic data for brevicompanine G (**5**): $C_{23}H_{29}N_3O_3$, $M = 395.49$, $a = 18.2192(5)$ Å, $b = 13.1497(3)$ Å, $c = 19.6255(5)$ Å, $\alpha = 90°$, $\beta = 116.2910(10)°$, $\gamma = 90°$, $V = 4215.44(19)$ Å^3, $T = 100(2)$ K, space group *P*21, $Z = 8$, μ(CuKα) = 0.668 mm^{-1}, 38,513 reflections measured, 14,878 independent reflections (R_{int} = 0.0360). The final R_1 values were 0.0364 ($I > 2\sigma(I)$). The final $wR(F^2)$ values were 0.0949 ($I > 2\sigma(I)$). The final R_1 values were 0.0366 (all data). The final $wR(F^2)$ values were 0.0952 (all data). The goodness of fit on F^2 was 1.045. Flack parameter = −0.03(4).

3.5. In Vitro IDH1(R132H) Inhibition Assay

The activity and inhibition of IDH1 (R132H) was determined by measuring the initial linear consumption of NADPH of the reaction. The enzyme activity assay was carried out in a 96-well microplate, using the purified IDH1 (R132H) protein in buffer solution containing 50 mM HEPES pH = 7.5, 4 mM $MgCl_2$, 100 mM NaCl, and 0.1 mg/mL bovine serum albumin. For inhibition assay, triplicate samples of compounds (10 μL) were incubated with the protein (2 μg/mL, 20 μL) and 60 μL buffer for 5 min. The reaction was initiated by adding 2 mM α-KG, 100 μM NADPH (10 μL) into the 96-well microplate. The consumption of NADPH was measured by monitoring the optical absorbance of each well every 30 s at 340 nm, which was the maximum absorption wavelength of NADPH, using a Biotek Synergy HT microplate reader.

3.6. Cytotoxicity Assay

Five human tumor cell lines (HL-60, MM231, A-549, HEP3B, and SW480), together with one non-cancerous cell line, the human normal colonic epithelial cell NCM460, were used in the cytotoxic activity assay. All cells were cultured in DMEM or RPMI-1640 medium (HyClone, Logan, UT, USA), supplemented with 10% fetal bovine serum (HyClone) at 37 °C in a humidified atmosphere with 5% CO_2. The cell survival assay was performed using the MTT method. Briefly, 100 μL suspended cells at an initial density of 1×10^5 cells/mL were seeded into each well of the 96-well culture plates and allowed to adhere for 12 h before addition of the test compounds. Each tumor cell line was exposed for 48 h to the test compounds at concentrations ranging from 0.0625 to 40 μM, with DDP (*cis*-platin, Sigma, St. Louis, MO, USA) and paclitaxel as positive controls. After incubation, culture supernatants were removed and exchanged with medium containing 0.5 mg/mL MTT. Then, after 4 h incubation in darkness at 37 °C, the medium was removed, and cells were added with 100 μL dimethyl sulfoxide. The absorbance at 570 nm was measured and data are expressed as averages of three replicates. The value of inhibition was calculated by using the following formula: % inhibition = $(1 - O_{treated}/OD_{control}) \times 100$. The IC_{50} values were calculated by using a standard dose-response curve fitting with Prism (version 5.0, GraphPad Software, La Jolla, CA, USA).

4. Conclusions

To sum up, four breviane spiroditerpenoids (**1–4**), including a new compound, brevione O (**1**), and three known compounds breviones I (**2**), J (**3**), and H (**4**), together with a known diketopiperazine alkaloid (**5**), were isolated and identified from an ethyl acetate extract of the fermented rice substrate of the coral-derived fungus *Penicillium* sp. TJ403-1. The absolute structure of **1** was elucidated on the basis of HRESIMS, one- and two-dimensional NMR spectroscopic data, and a comparison of its ECD spectrum with data gathered from the literature. Remarkably, despite the fact that the structure of brevicompanine G (**5**) was once established via a combination of NMR spectroscopic data and biosynthetic logic-based consideration, our current work is the first to confirm the absolute configuration of **5** by single-crystal X-ray crystallography, which will be of great significance for the structure elucidation of complex diketopiperazine alkaloids. Remarkably, compound **2** showed significant inhibitory activities against HL-60, A-549, and HEP3B tumor cell lines, with IC_{50} values of 4.92 ± 0.65, 8.60 ± 1.36, and 5.50 ± 0.67 μM, respectively.

Supplementary Materials: The following are available online at http://www.mdpi.com/1660-3397/16/4/110/s1, 1D (^{1}H- and ^{13}C-) and 2D (HMBC, HSQC, COSY, NOESY) NMR, HRESIMS, IR, and UV spectra of **1**.

Acknowledgments: We thank the Analytical and Testing Center at HUST for the ECD and IR analyses. The support from the Program for Changjiang Scholars of Ministry of Education of the People's Republic of China (No. T2016088), the Innovative Research Groups of the National Natural Science Foundation of China (No. 81721005), the National Science Fund for Distinguished Young Scholars (No. 8172500151), the China Postdoctoral Science Foundation Funded Project (No. 2017M610479), the National Natural Science Foundation of China (Nos. 81573316, 21702067), the Academic Frontier Youth Team of HUST, and the Integrated Innovative Team for Major Human Diseases Program of Tongji Medical College (HUST) are greatly acknowledged.

Author Contributions: Beiye Yang contributed to the extraction, isolation, identification, and manuscript preparation. Weiguang Sun contributed to the bioactivity tests. Shuang Lin contributed to the fungal isolation and fermentation. Jianping Wang advised and assisted Yang's experiments. Xiao-Nian Li contributed to the X-ray diffraction experiment. Hucheng Zhu and Zengwei Luo contributed to the structure elucidation of isolated compounds and shared in the tasks of the manuscript preparation. Yongbo Xue contributed to the NMR experiments. Zhengxi Hu guided the experiments and wrote the manuscript. Yonghui Zhang designed the experiments and revised the manuscript.

Conflicts of Interest: The authors declare no conflict of interest.

References

1. Sueyoshi, K.; Yamano, A.; Ozaki, K.; Sumimoto, S.; Iwasaki, A.; Suenaga, K.; Teruya, T. Three new malyngamides from the marine cyanobacterium *Moorea producens*. *Mar. Drugs* **2017**, *15*, 367. [CrossRef] [PubMed]
2. Imhoff, J.F. Natural products from marine fungi—Still an underrepresented resource. *Mar. Drugs* **2016**, *14*, 19. [CrossRef] [PubMed]
3. Mayer, A.M.S.; Glaser, K.B.; Cuevas, C.; Jacobs, R.S.; Kem, W.; Little, R.D.; McIntosh, J.M.; Newman, D.J.; Potts, B.C.; Shuster, D.E. The odyssey of marine pharmaceuticals: A current pipeline perspective. *Trends Pharmacol. Sci.* **2010**, *31*, 255–265. [CrossRef] [PubMed]
4. Newman, D.J.; Cragg, G.M. Current Status of marine-derived compounds as warheads in anti-tumor drug candidates. *Mar. Drugs* **2017**, *15*, 99. [CrossRef] [PubMed]
5. Cherigo, L.; Lopez, D.; Martinez-Luis, S. Marine natural products as breast cancer resistance protein inhibitors. *Mar. Drugs* **2015**, *13*, 2010–2029. [CrossRef] [PubMed]
6. Geris, R.; Simpson, T.J. Meroterpenoids produced by fungi. *Nat. Prod. Rep.* **2009**, *26*, 1063–1094. [CrossRef] [PubMed]
7. Macías, F.A.; Varela, R.M.; Simonet, A.M.; Cutler, H.G.; Cutler, S.J.; Ross, S.A.; Dunbar, D.C.; Dugan, F.M.; Hill, R.A. (+)-Brevione A. The first member of a novel family of bioactive spiroditerpenoids isolated from *Penicillium brevicompactum* Dierckx. *Tetrahedron Lett.* **2000**, *41*, 2683–2686. [CrossRef]
8. Macías, F.A.; Varela, R.M.; Simonet, A.M.; Cutler, H.G.; Cutler, S.J.; Dugan, F.M.; Hill, R.A. Novel bioactive breviane spiroditerpenoids from *Penicillium brevicompactum* Dierckx. *J. Org. Chem.* **2000**, *65*, 9039–9046. [CrossRef] [PubMed]
9. Li, Y.; Ye, D.; Chen, X.; Lu, X.; Shao, Z.; Zhang, H.; Che, Y. Breviane spiroditerpenoids from an extreme-tolerant *Penicillium* sp. isolated from a deep sea sediment sample. *J. Nat. Prod.* **2009**, *72*, 912–916. [CrossRef] [PubMed]
10. Li, Y.; Ye, D.; Shao, Z.; Cui, C.; Che, Y. A sterol and spiroditerpenoids from a *Penicillium* sp. isolated from a deep sea sediment sample. *Mar. Drugs* **2012**, *10*, 497–508. [CrossRef] [PubMed]
11. Kwon, J.; Seo, Y.H.; Lee, J.E.; Seo, E.K.; Li, S.; Guo, Y.; Hong, S.B.; Park, S.Y.; Lee, D. Spiroindole alkaloids and spiroditerpenoids from *Aspergillus duricaulis* and their potential neuroprotective effects. *J. Nat. Prod.* **2015**, *78*, 2572–2579. [CrossRef] [PubMed]
12. Yokoe, H.; Mitsuhashi, C.; Matsuoka, Y.; Yoshimura, T.; Yoshida, M.; Shishido, K. Enantiocontrolled total syntheses of breviones A, B, and C. *J. Am. Chem. Soc.* **2011**, *133*, 8854–8857. [CrossRef] [PubMed]
13. Macías, F.A.; Carrera, C.; Chinchilla, N.; Fronczek, F.R.; Galindo, J.C.G. Synthesis of the western half of breviones C, D, F and G. *Tetrahedron* **2010**, *66*, 4125–4132. [CrossRef]
14. Takikawa, H.; Hirooka, M.; Sasaki, M. The first synthesis of (±)-brevione B, an allelopathic agent isolated from *Penicillium* sp. *Tetrahedron Lett.* **2003**, *44*, 5235–5238. [CrossRef]
15. Hu, Z.X.; Xue, Y.B.; Bi, X.B.; Zhang, J.W.; Luo, Z.W.; Li, X.N.; Yao, G.M.; Wang, J.P.; Zhang, Y.H. Five new secondary metabolites produced by a marine-associated fungus, *Daldinia eschscholzii*. *Mar. Drugs* **2014**, *12*, 5563–5575. [CrossRef] [PubMed]
16. Du, L.; Yang, X.; Zhu, T.; Wang, F.; Xiao, X.; Park, H.; Gu, Q. Diketopiperazine alkaloids from a deep ocean sediment derived fungus *Penicillium* sp. *Chem. Pharm. Bull.* **2009**, *57*, 873–876. [CrossRef] [PubMed]
17. He, Y.; Zheng, M.; Li, Q.; Hu, Z.; Zhu, H.; Liu, J.; Wang, J.; Xue, Y.; Li, H.; Zhang, Y. Asperspiropene A, a novel fungal metabolite as an inhibitor of cancer-associated mutant isocitrate dehydrogenase 1. *Org. Chem. Front.* **2017**, *4*, 1137–1144. [CrossRef]

18. Sun, X.; Wang, W.; Chen, J.; Cai, X.; Yang, J.; Yang, Y.; Yan, H.; Cheng, X.; Ye, J.; Lu, W.; et al. The natural diterpenoid isoforretin A inhibits thioredoxin-1 and triggers potent ROS-mediated antitumor effects. *Cancer Res.* **2017**, *77*, 926–936. [CrossRef] [PubMed]
19. Hu, Z.X.; Liu, M.; Wang, W.G.; Li, X.N.; Hu, K.; Li, X.R.; Du, X.; Zhang, Y.H.; Puno, P.T.; Sun, H.D. 7α,20-Epoxy-*ent*-kaurane diterpenoids from the aerial parts of *Isodon pharicus*. *J. Nat. Prod.* **2018**, *81*, 106–116. [CrossRef] [PubMed]

![marine drugs logo] *marine drugs*

MDPI

Article

The Maxi-K (BK) Channel Antagonist Penitrem A as a Novel Breast Cancer-Targeted Therapeutic

Amira A. Goda [1], Abu Bakar Siddique [1], Mohamed Mohyeldin [1,3], Nehad M. Ayoub [2] and Khalid A. El Sayed [1,*]

[1] Department of Basic Pharmaceutical Sciences, School of Pharmacy, University of Louisiana at Monroe, 1800 Bienville Drive, Monroe, LA 71201, USA; amirakareem16@gmail.com (A.A.G.); siddiqab@warhawks.ulm.edu (A.B.S.); Mohamed.mohyeldin@alexu.edu.eg (M.M.)
[2] Department of Clinical Pharmacy, Faculty of Pharmacy, Jordan University of Science and Technology, Irbid 22110, Jordan; nmayoub@just.edu.jo
[3] Department of Pharmacognosy, Faculty of Pharmacy, Alexandria University, Alexandria 21521, Egypt
*   Correspondence: elsayed@ulm.edu; Tel.: +1-318-342-1725

Received: 6 April 2018; Accepted: 9 May 2018; Published: 11 May 2018

Abstract: Breast cancer (BC) is a heterogeneous disease with different molecular subtypes. The high conductance calcium-activated potassium channels (BK, Maxi-K channels) play an important role in the survival of some BC phenotypes, via membrane hyperpolarization and regulation of cell cycle. BK channels have been implicated in BC cell proliferation and invasion. Penitrems are indole diterpene alkaloids produced by various terrestrial and marine *Penicillium* species. Penitrem A (**1**) is a selective BK channel antagonist with reported antiproliferative and anti-invasive activities against multiple malignancies, including BC. This study reports the high expression of BK channel in different BC subtypes. In silico BK channel binding affinity correlates with the antiproliferative activities of selected penitrem analogs. **1** showed the best binding fitting at multiple BK channel crystal structures, targeting the calcium-sensing aspartic acid moieties at the calcium bowel and calcium binding sites. Further, **1** reduced the levels of BK channel expression and increased expression of TNF-α in different BC cell types. Penitrem A (**1**) induced G1 cell cycle arrest of BC cells, and induced upregulation of the arrest protein p27. Combination treatment of **1** with targeted anti-HER drugs resulted in synergistic antiproliferative activity, which was associated with reduced EGFR and HER2 receptor activation, as well as reduced active forms of AKT and STAT3. Collectively, the BK channel antagonists represented by penitrem A can be novel sensitizing, chemotherapeutics synergizing, and therapeutic agents for targeted BC therapy.

Keywords: breast cancer; BK (Maxi-K) channel; EGFR; HER2; penitrem A; gefitinib; lapatinib; TNF-α

1. Introduction

Breast cancer (BC) is the most commonly diagnosed malignancy among women, globally contributing to high mortality rates [1]. BC is highly heterogeneous at both molecular and clinical level, with various pathological and molecular subtypes [2]. Four major molecular subtypes of BC were identified; these include luminal A, luminal B, human epidermal growth factor receptor 2 (HER2)-positive, and basal-like subtypes [3–5]. Both luminal A and luminal B tumors are hormone receptor-positive, and have expression patterns reminiscent of the luminal epithelial component of the breast [6]. HER2-enriched tumors are characterized by overexpression/amplification of the ErbB2/HER2 gene, and are generally hormone receptor-negative [2,7]. Basal-like tumors are predominantly triple-negative, lacking the expression of hormone receptors and HER2 [7]. The subtypes are associated with distinct pathological features and clinical outcomes [2].

The Maxi-K channels, also known as BK/BKCa/Slo1/KCa1.1 channels, are characterized by ubiquitous tissue expression and large conductance to potassium ions [8]. The channel is sensitive to voltage and Ca^{2+}. BK channels are essential for smooth muscle contraction, neurotransmitter release, hormone secretion, and gene expression [9]. In addition, BK channels have been shown to associate with multiplicity of plasma membranes and intracellular proteins as linkers of membrane potential, cell metabolism, and cell signaling [8]. The structure of BK channel is composed of four Slo1 subunits. Each Slo1 subunit consists of four α-subunits and four β-subunits [10–12]. Each α-subunit is composed of seven putative transmembrane-spanning α-helical segments (pore-gate domain, PGD). It controls ions conductance, selectivity, and voltage change sensing [10–12]. The cytoplasmic carboxy tail of the α-subunit possesses two intrinsic high-affinity Ca^{2+} binding and phosphorylation sites, known to direct the gating mechanism (cytosolic tail domain, CTD). Each β-subunit is formed of two transmembrane domains with a long extracellular linker, and both its amino- and carboxy-terminals are cytoplasmic (voltage sensor domain, VSD) [10–12]. Dysregulation or overexpression of BK channels have been associated with altered cell cycle progression [13], cell proliferation [14–20], and migration. These features are fundamental for cancer development and progression [21–24]. In this regard, earlier studies indicated variable levels of BK channel expression in human BC cells [25]. In addition, electrophysiological studies on cervical and BC cells suggested that BK channels are directly activated by estrogens, which could have an essential role in uterus, breast, and prostate cancers [26,27]. Taken together, pharmacological blockade of BK channels would be expected to suppress cancer cell proliferation [13].

Penitrems are indole diterpene neurotoxic alkaloids produced by various terrestrial and marine *Penicillium* species [28,29]. This study team reported penitrems **1**, **2** (Scheme 1), and others from a marine-derived *Penicillium commune* isolate GS20 isolated from sponge and sediment samples collected in the Arabian Gulf [30,31]. Penitrems have potent tremorgenic activity in mammals, secondary to the antagonism of BK channels [28,29]. Previous findings from our laboratory revealed the potential anticancer effects of penitrems as inhibitors of proliferation, migration, and invasion of BC cells [30,31]. The mechanism for these reported anticancer effects was associated with the suppression of the Wnt/β-catenin pathway in BC cells [30].

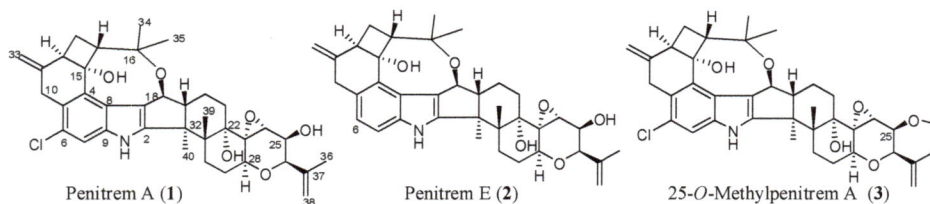

Scheme 1. Chemical structures of tested penitrems 1–3.

In this study, penitrems were applied in terms of BK channel inhibitors, to assess their antiproliferative effects in multiple BC cell lines, in vitro. The antiproliferative activity of the most potent **1** was assessed individually, and in combination with targeted therapy. The study also compares the in silico binding mode of **1** at multiple BK channel crystal structures with its related less active analogs, **2** and **3** (Scheme 1).

2. Results

2.1. Antiproliferative Effects of Penitrems in Breast Cancer Cells In Vitro

The antiproliferative activity of penitrems was assessed using MTT cell viability assay. Multiple human BC cell lines representing the different molecular subtypes were tested, including MDA-MB-231, BT-474, and SK-BR-3 cells, along with the human neuronal Schwann cells CRL-2765 and the non-tumorigenic mammary epithelial MCF-12A cells. Penitrem A (**1**) resulted in a dose-dependent inhibition among all three tested BC cell lines after 48 h culture duration (Figure 1). Among BC cell lines exposed to **1**, the triple-negative MDA-MB-231 cells were most sensitive to the antiproliferative effects of **1**, as indicated by lowest IC_{50} value (Table 1). Penitrem E (**2**) and 25-*O*-methylpenitrem A (**3**) showed less inhibition of BC cell growth compared to **1**. MDA-MB-231 cells were the most sensitive to growth suppression by **2**, while the HER2-positive SK-BR-3 cells were most inhibited by **3**. With respect to non-tumorigenic cells, **1** was the most toxic, with IC_{50} values of 33.7 μM in MCF-12A (non-tumorigenic human mammary epithelial cells), and 22.6 μM in Schwann C RL-2765 (peripheral neuronal cells), respectively (Figure 1, Table 1).

Table 1. The IC_{50} (μM) values of penitrems **1**–**3** in multiple breast cancer (BC) and non-cancer cell lines in vitro.

Compound	MCF-12A	CRL-2765	MDA-MB-231	BT-474	SK-BR-3
1	33.7	22.6	9.8	10.3	15.1
2	44.0	67.8	20.3	31.8	36.7
3	78.8	48.2	37.8	22.4	27.1

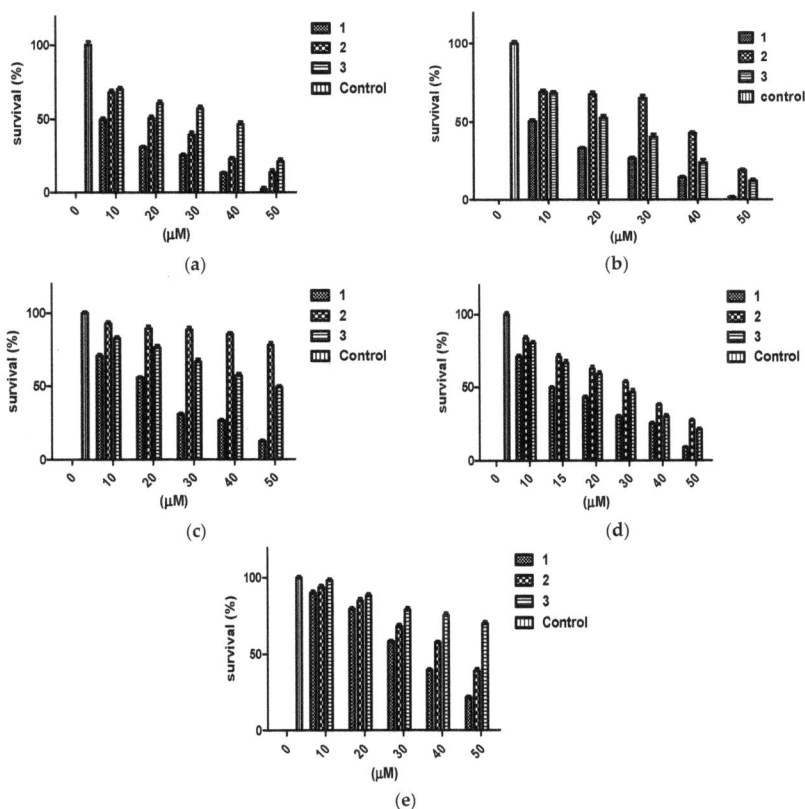

Figure 1. Antiproliferative effects of penitrems **1**–**3** against BC and non-tumorigenic cells. Viability of (**a**) BT-474, (**b**) MDA-MB-231, (**c**) SK-BR-3, (**d**) CRL-2765, and (**e**) MCF-12A cells after incubation with increasing concentrations of **1**–**3** for 48 h in cell cultures. All values are relative to vehicle control (100% survival). Bars represent mean ± SEM for four replicates within each group.

2.2. In Silico Binding of Penitrems with BK Channel

Both human BK channel PDB crystal structures, 3NAF and 3MT5, were used in this study for molecular docking [32–34]. Penitrems **1–3** were virtually screened for their ability to bind the BK channel crystal structures 3NAF and 3MT5, to correlate their binding modes and affinity with the antiproliferative activity of the compounds. Binding of the intracellular Ca^{2+} at the CTD enhances the BK channel opening. Each Slo1 subunit CTD has two Ca^{2+} high affinity binding sites; the RCK1, which includes the side-chain carboxylates of Asp367 and Glu535, as well as the main chain carbonyl of Arg514, in addition to the C-terminus of RCK2 domain, with a string of aspartic residues at the calcium bowel [35]. In the calcium bowel, the side chain carboxylate groups of Asp895 and Asp897 provide direct coordinates to bind Ca^{2+} ion, consistent with previous mutagenesis and biochemistry literature [36]. The side chain of Asp894 does not directly contact with Ca^{2+}, but instead, it forms salt bridges with Arg1018 and Lys1030. The N-terminus of the RCK1 domain identified two acidic residues, GLUlu374 and Glu399, which are close to the voltage sensor activation site Asp99 and Asn172 at the edge of VSD.

In PDB 3MT5, the C-15 tertiary alcohol group of **1** contributed a hydrogen bonding donor interaction with Asp367, which has high affinity for Ca^{2+} binding (Figure 2). Similarly, the C-25 secondary alcohol of **1** contributed hydrogen bonding donor interaction with Asp895, and accepted hydrogen bonding interaction with Asp897 at the Ca^{2+} bowel site, which provides direct coordinate to bind Ca^{2+} (Figure 3). In PDB structure 3NAF, the NH-1 of **1** contributed hydrogen bonding donor interaction with Asp894 and the C-25 hydroxyl group contributed hydrogen bonding donor interaction with Asp895 at the Ca^{2+} bowel site, justifying its antagonistic activity to BK channel (Figure 4). Meanwhile, penitrem analog **3** showed only two possible interactions at the calcium bowel of PDB 3MT5 (Figure S1). Its C-15 tertiary hydroxyl group showed hydrogen bonding donor interaction with Glu521, while its NH-1 showed hydrogen bonding donor interaction with Gln525.

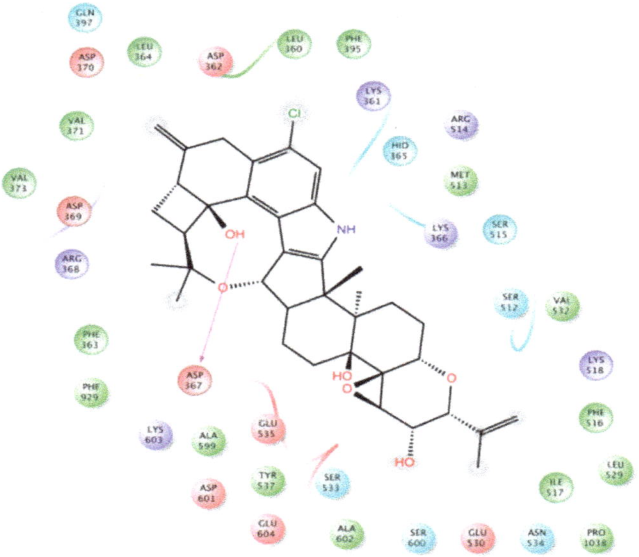

(**a**)

Figure 2. *Cont.*

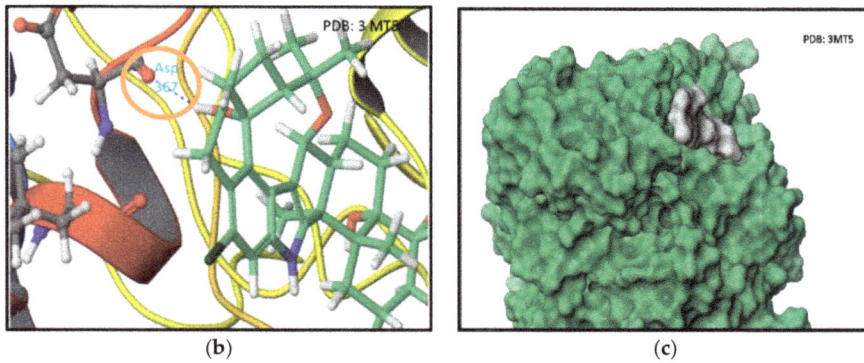

Figure 2. Binding mode of **1** at the Ca^{2+} binding site of BK channel crystal structure PDB 3MT5. (**a**) The 2D binding mode of **1** at the Ca^{2+} binding site. Its C-15 hydroxyl group contributed hydrogen bonding donor interaction with Asp367, which has high affinity for the Ca^{2+} binding. (**b**) The overlay of 3D structure of **1** at the Ca^{2+} binding site of the PDB 3MT5. (**c**) Shape fitting of **1** within the Ca^{2+} binding site of BK PDB structure 3MT5.

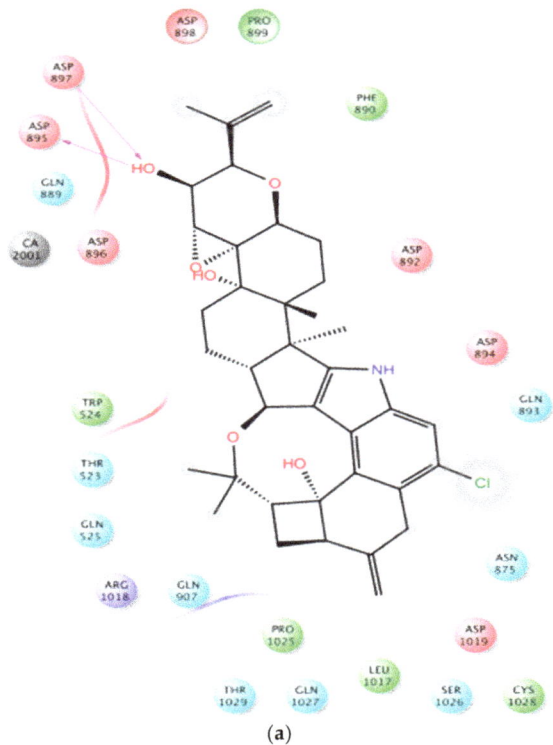

Figure 3. *Cont.*

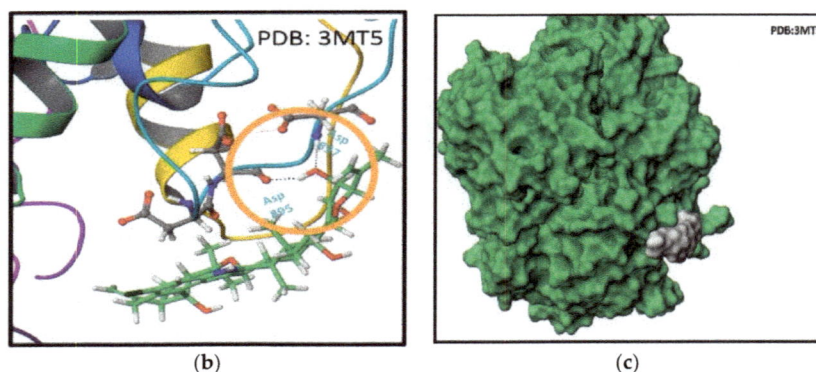

Figure 3. Binding mode of **1** at the Ca^{2+} binding bowel of the BK channel crystal structure PDB 3MT5. (**a**) The 2D binding mode of **1** at the Ca^{2+} binding bowel. Its C-25 hydroxyl group contributed hydrogen bonding donor interaction with ASP895 and accepted hydrogen bonding interaction with Asp897. Both aspartic acid moieties are known to provide direct coordinate for Ca^{2+} binding at the Ca^{2+} bowel site. (**b**) The overlay of 3D structure of **1** at the Ca^{2+} binding bowel of the PDB 3MT5. (**c**) Shape fitting of **1** within the Ca^{2+} binding bowel pocket of BK channel PDB structure 3MT5.

There were no interactions with the essential aspartates at the calcium bowel, which justify its weak BK channel inhibitory activity. The only important interaction for **3** in the calcium bowel of PDB 3NAF was a hydrogen bonding donor interaction between its NH-1 with Asp892 (Figure 5). Penitrem **2** contributed only a hydrogen bonding donor interaction via its NH-1 with Phe890 at the calcium bowel of PDB 3NAF, and showed no interactions with PDB 3MT5 (Figures S2 and S3).

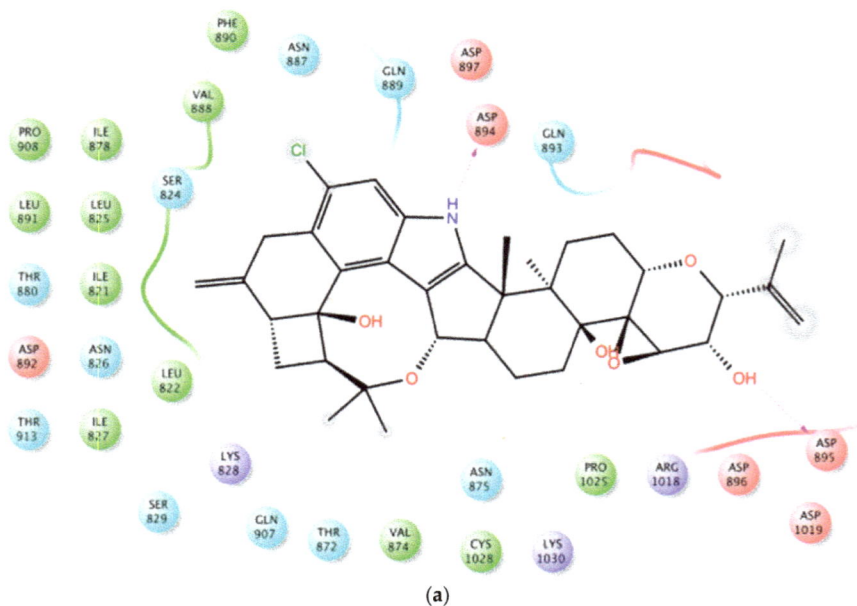

Figure 4. *Cont.*

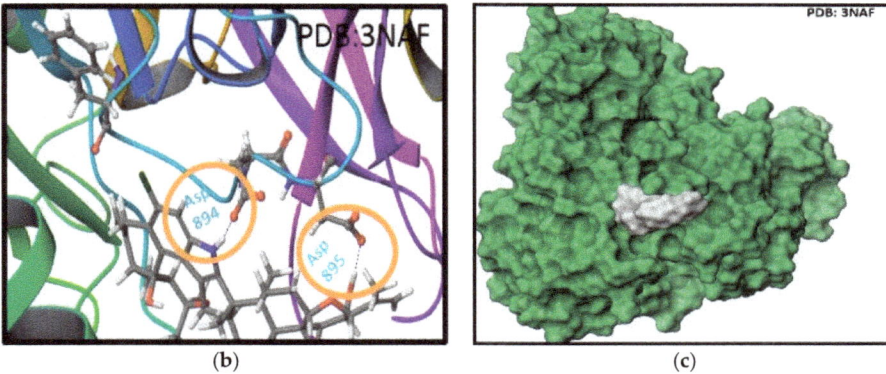

Figure 4. Binding mode of **1** at the BK channel crystal structure PDB 3NAF. (**a**) The 2D binding mode of **1** at the Ca^{2+} binding bowel of PDB 3NAF. NH-1 of **1** contributed hydrogen bonding donor interaction with Asp894 while the C-25 hydroxyl contributed hydrogen bonding donor interaction with Asp895 at the calcium bowel site. These aspartic acid moieties provide direct coordinate for Ca^{2+} binding at the Ca^{2+} bowel site. (**b**) The overlay of 3D structure of **1** at the Ca^{2+} binding bowel of the PDB 3NAF. (**c**) Shape fitting of **1** within the Ca^{2+} binding bowel pocket PDB structure 3NAF.

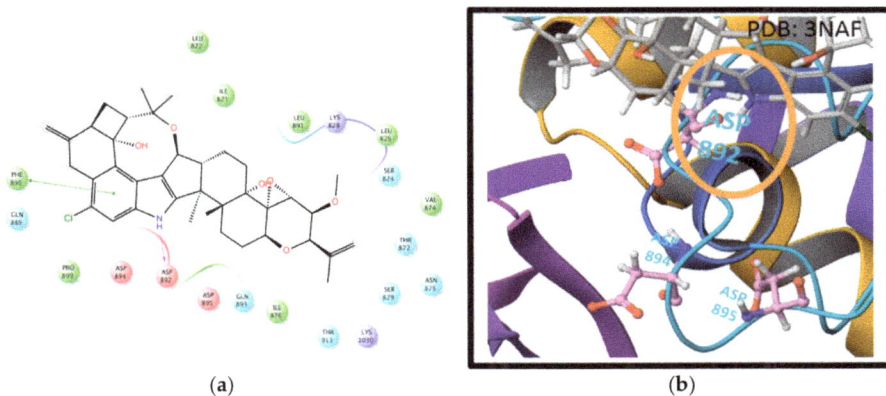

Figure 5. Binding mode of **3** at the Ca^{2+} bowel of the PDB crystal structure 3NAF. (**a**) The 2D binding mode of **3** at the Ca^{2+} bowel showing its NH-1 hydrogen bonding donor interaction with the critical Asp892. (**b**) The overlay of 3D structure of **3** at the calcium bowel of the PDB 3NAF crystal structure.

2.3. Expression of BK Channels in BC Cells and In Vitro Impact of Penitrems on Channel Expression

The basic functional unit of BK channels is the tetramer of the pore-forming α-subunits (KCa1.1 or Slo1) encoded by the gene KCNMA1 [37–39]. The expression levels of BK channel subunits α-1 (KCNMA1) in multiple BC cell lines was compared with the neuronal Schwann cells, CRL-2765, and the non-tumorigenic mammary epithelial MCF-12A cells using Western blot analysis (Figure 6).

The effect of penitrems on the levels of KCNMA1 and TNF-α, a marker for BK antagonism [40–45] were assessed in BC cell lines (Figures 7–9). Consistent with its antiproliferative activity, **1** resulted in the greatest reduction in the total levels of BK channel subunits α-1 (KCNMA1), an effect which was also associated with increased total levels of TNF-α among BC cell lines (Figures 7–9). Similar effects were observed with 25-*O*-methylpenitrem A (**3**), however, to a lesser extent than **1**. Penitrem **2** treatment did not cause changes to the total levels of KCNMA1 in BC cells. Compound **2**, however, increased total levels of TNF-α in both BT-474 and SK-BR-3 cells compared to control groups (Figures 7–9).

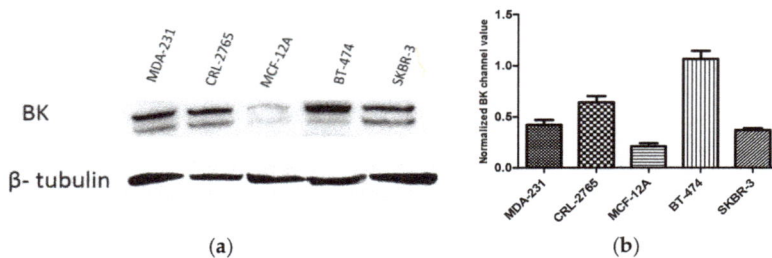

(a)　　　　　　　　　　(b)

Figure 6. Expression of BK channel subunits α-1 (KCNMA1) in BC and non-tumorigenic cell lines. (a) Western blots representing total levels of the BK channel subunits α-1 (KCNMA1) in the different cell lines. (b) Bar graphs indicating densitometric quantitative analysis performed on all blots, in which optical density of each band was normalized with corresponding β-tubulin.

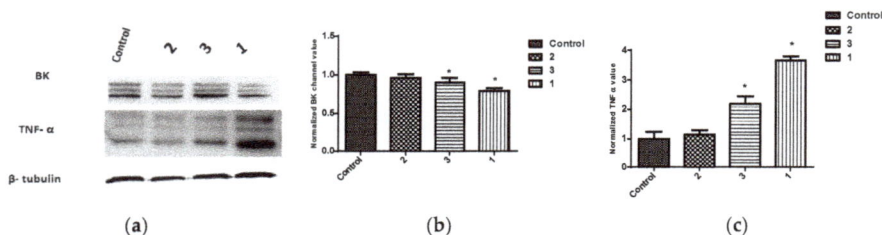

(a)　　　　　　　　(b)　　　　　　　　(c)

Figure 7. In vitro effects of 10 μM treatments of penitrems **1–3** on the expression of BK channel (KCNMA1) and TNF-α (D2D4) in MDA-MB-231 cells using Western blot analysis. (a) Western blots for cells treated with penitrems **1–3**. (b) Western blot quantification of the in vitro effects of penitrem **1–3** treatments on the expression of KCNMA1. (c) Western blotting quantification of the effects of **1–3** treatments on the activation of TNF-α. Vertical bars indicate the normalized protein value ± SEM. *: indicate significant differences ($p \leq 0.05$).

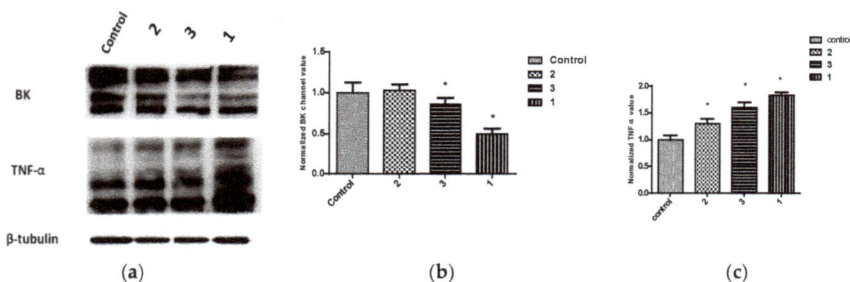

(a)　　　　　　　　(b)　　　　　　　　(c)

Figure 8. In vitro effects of 10 μM treatments of penitrems **1–3** on the expression of BK channel (KCNMA1) and activation of TNF-α (D2D4) in BT-474 cells using Western blot analysis. (a) Western blot for cells treated with penitrems **1–3**. (b) Western blot quantification of the in vitro effects of penitrems **1–3** treatment on the expression of KCNMA1. (c) Western blot quantification of the effects of penitrems **1–3** treatment on the activation of TNF-α. Vertical bars indicate the normalized protein value ± SEM. *: indicate significant differences ($p \leq 0.05$).

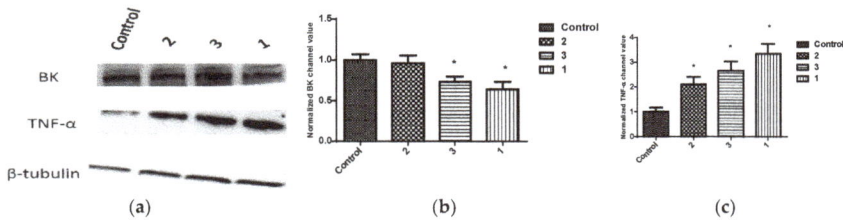

Figure 9. In vitro effects of 10 μM treatments of penitrems **1–3** on the expression of BK channel (KCNMA1) and activation of TNF-α (D2D4) in SK-BR-3 cells using Western blot analysis. (**a**) Western blot for cells treated with penitrems **1–3**. (**b**) Western blotting quantification of the in vitro effects of penitrem **1–3** treatments on the expression of KCNMA1. (**c**) Western blot quantification of the effects of **1–3** treatments on the activation of TNF-α. Vertical bars indicate the normalized protein value ± SEM. *: indicate significant differences ($p \leq 0.05$).

In the same context, immunofluorescent staining of MDA-MB-231 (Figure 10a,b) and BT-474 cells (Figure 10c,d) indicated strong cytoplasmic expression of KCNMA1 in vehicle-treated culture media (Figure 10a,c). Penitrem **1** treatment caused significant reduction in the total level of KCNMA1 compared to cells in vehicle-treated control groups (Figure 10a,c). Penitrem **1** treatments caused remarkable reduction in the total levels of KCNMA1 in both cell lines compared to cells of vehicle-treated controls (Figure 10b,d).

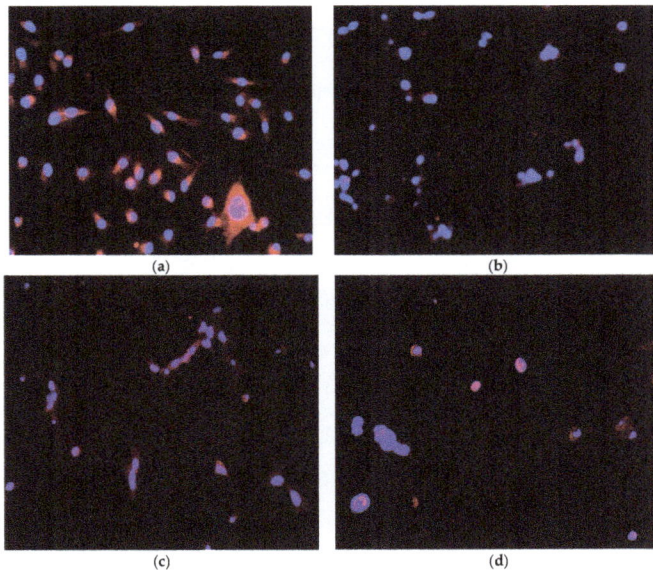

Figure 10. Immunocytochemical fluorescence staining of the total levels of BK channel subunits α-1 (KCNMA1) in MDA-MB-231 and BT-474 BC cells treated with **1** at its IC_{50} concentration, 9.8 and 10.3 μM, respectively, for 24 h. (**a**) MDA-MB-231 cells treated with vehicle control. (**b**) MDA-MB-231 cells treated with **1** at 9.8 μM. (**c**) BT-474 cells treated with vehicle control. (**d**) BT-474 cells treated with **1** at 10.3 μM. Red staining indicates positive immunofluorescence signal for KCNMA1 and blue staining indicates cell nuclei counter-stained with DAPI. Magnification of each photomicrograph is 20×. (**a**) or treatment media containing 1 treatment for 24 h (**b**). Cells were then fixed in pre-cooled acetone and subjected to immunofluorescence analysis for detection of total KCNMA1 level.

2.4. Effect of Penitrem A Treatment on Cell Cycle Progression in BC Cells

The effect of **1** on cell cycle progression of BT-474 cells was evaluated using flow cytometry by applying propidium iodide (PI) staining (Figure 11) [13,45–48]. BT-474 cells were treated with various doses of **1** for 48 h, prior to fixation and staining. BT-474 cells exposed to various concentrations of **1** resulted in a dose-dependent increase (53% to 70%) in the proportion of cells in G1 phase compared to cells in vehicle-treated control group, with the maximal effect observed at 30 μM.

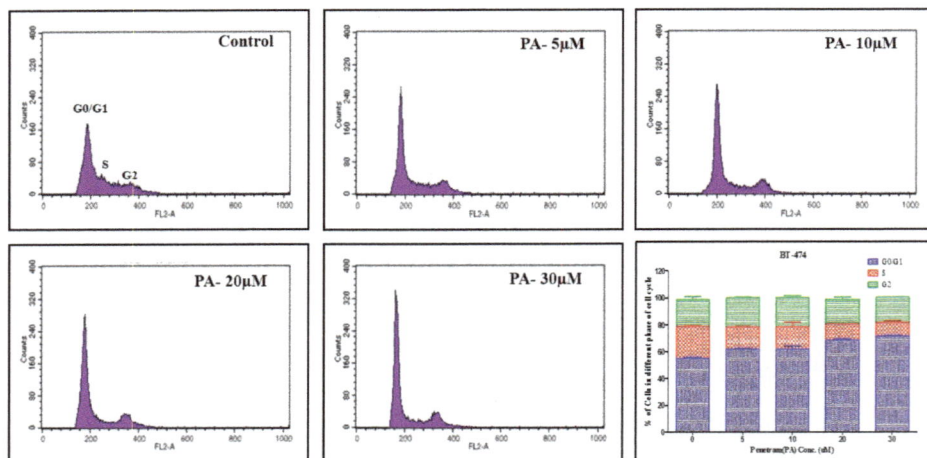

Figure 11. Flow cytometry analysis for cell cycle progression in BT-474 cells treated with **1** (PA). Cells in the various treatment groups were synchronized in G1 phase. Histograms were generated using Cell Quest software (PI staining). Column graph shows percentage of BT-474 cells in each phase of the cell cycle. Vertical bars show the average percentage of three independent experiments.

2.5. Effects of Combined Treatment of Targeted Agents and Penitrem A on the Growth of BC Cells

Penitrem **1** is known to be a potent BK channels antagonist at very low molar concentration [49,50]. Therefore, subeffective dose combinations of **1** with either lapatinib (LP) or gefitinib (GF) were hypothesized to have a superior antiproliferative activity, compared to individual treatments in BC cells.

LP treatment inhibited the growth of BT-474 cells in a dose-dependent manner, with an IC_{50} value of 123.0 nM (data not shown). Penitrem **1** has been shown to inhibit growth of BT-474 cells at an IC_{50} value of 10.3 μM (Table 1). Figure 12 shows the effect of a combined range of subeffective concentrations of LP (10–80 nM) with a range of **1** concentrations (0.5, 1.0, 2.5, and 5.0 μM). Combined treatment of LP and **1** resulted in dose-dependent inhibition of BT-474 cell proliferation after 48 h in culture (Figure 12A). The combination of LP with **1** effectively reduced the IC_{50} concentration of LP within each treatment combination, compared to individual monotherapy treatments (Table 2).

Table 2. Summary of combinations of **1** with lapatinib (LP) or gefitinib (GF) against BT-474 BC cells.

IC_{50}/Combination Index (CI) Values				
LP	LP + 1 (0.5 μM)	LP + 1 (1.0 μM)	LP + 1 (2.5 μM)	LP + 1 (5.0 μM)
123.0 nM	65.68 nM/0.58	73.56 nM/0.70	69.80 nM/0.81	66.17 nM/1.03
GF	GF + 1 (0.5 μM)	GF + 1 (1.0 μM)	GF + 1 (2.5 μM)	GF + 1 (5.0 μM)
302.4 nM	124.77 nM/0.46	176.78 nM/0.68	130.84 nM/0.68	114.42 nM/0.87

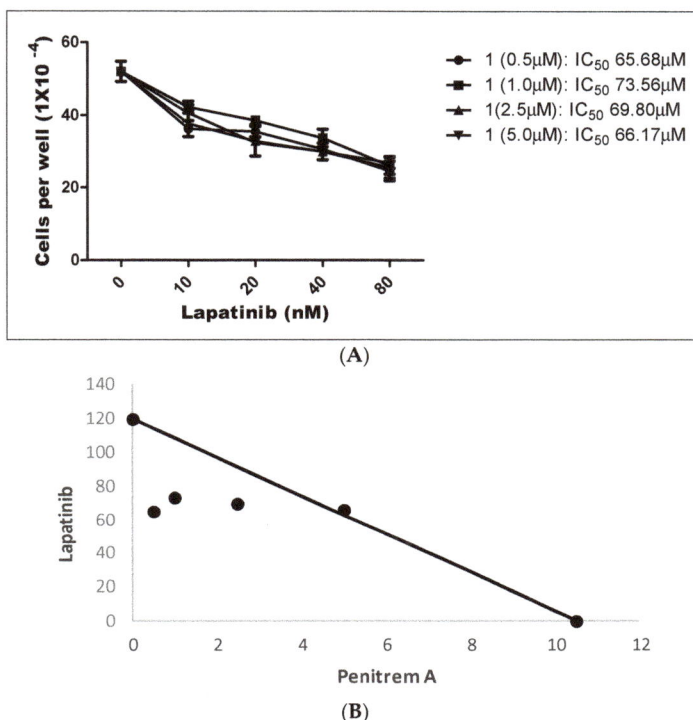

Figure 12. (**A**) Effects of combined treatment of penitrem **1** and LP on BT-474 cell viability after 48 h culture period. Bars represent mean ± SEM for four replicates within each group. (**B**) Isobologram of penitrem A (**1**, μM) and LP (nM) antiproliferative effect in BT-474 cells. IC_{50} concentrations for **1** and LP were plotted on the *X*- and *Y*-axis, respectively. The solid line connecting these points represents the concentration of each compound required to induce the same relative growth inhibition when used in combination, if the interaction between the compounds is additive. Each data point on the isobologram represents the actual concentrations of **1** and LP which induced 50% inhibition of cell growth when used in combination.

To further evaluate the nature of interaction between both compounds, combination index (CI) and isobologram analysis were conducted. CI is a quantitative representation of pharmacological interaction between two drugs, in which values <1, 1, and >1 indicate synergistic, additive, and antagonistic effects, respectively [51,52]. CI values are calculated as follows: $CI = [Xc/X + Tc/T]$, where X and T stand for the concentrations of individual combination ingredients that induce 50% cell growth inhibition (IC_{50}); Xc and Tc are the concentrations of combination ingredients that induce 50% cell growth inhibition when used in combination as determined by non-linear regression curve fit analysis [51,52]. Isobologram analysis is a graphical method used to evaluate the effect of equally effective dose pairs for a single effect level [53]. Calculated CI of **1**-LP combined treatments indicated synergism with values of 0.58, 0.70, and 0.81, respectively, for the first three treatments (Figure 12A,B and Table 2). The use of combined 5.0 μM of **1** with LP afforded an IC_{50} of 66.17 nM and CI of 1.03, a very weakly antagonistic combination (Figure 12A,B and Table 2). In addition, isobologram showed synergistic interaction between both compounds as indicated by three combinations of data points located below the line of additive activity (Figure 12B).

Gefitinib (GF), an EGFR small molecule inhibitor, suppressed the growth of BT-474 cells in a dose-dependent fashion, with IC_{50} value of 302.4 nM (data not shown). Figure 13A shows the effects of combined subeffective concentrations of GF (50–200 nM) with range of **1** concentrations (0.5, 1.0, 2.5, and 5.0 µM). The combination of both GF and **1** suppressed the growth of BT-474 cells in a dose-dependent manner (Figure 13A). The combination of GF with **1** effectively reduced the GF IC_{50} concentrations for each tested treatment combination, compared to single GF treatment (Table 2). Calculated CI of **1**-GF combined treatments indicated synergism, with values of 0.46, 0.68, 0.68, and 0.87, respectively (Figure 13A,B and Table 2). Isobologram analysis showed synergistic interaction between both compounds, as indicated by four data points located below line of additive activity (Figure 13B).

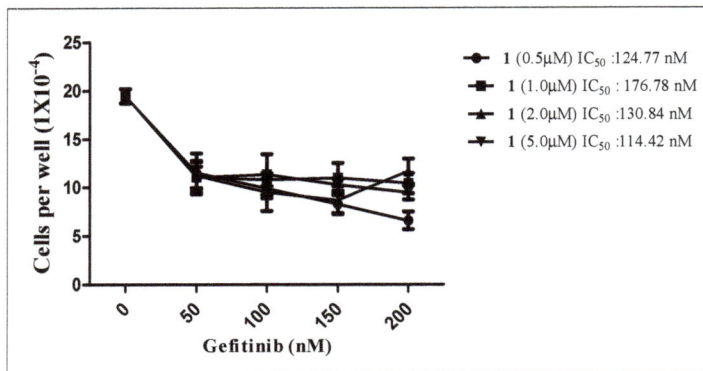

(A)

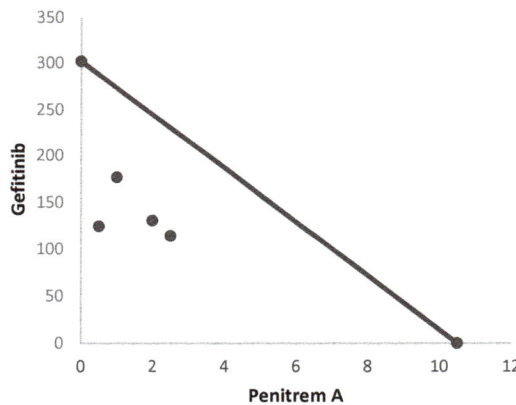

(B)

Figure 13. (**A**) Effects of combined subeffective doses of **1** and GF in BT-474 cells after 48 h culture period. Bars represent mean ± SEM for four replicates within each group. (**B**) Isobologram of **1** (µM) and GF (nM) antiproliferative effect in BT-474 cells. IC_{50} concentrations for **1** and GF were plotted on the *X*- and *Y*-axis, respectively. The solid line connecting these points represents the concentration of each compound required to induce the same relative growth inhibition when used in combination, if the interaction between the compounds is additive. Each data point on the isobologram represents the actual concentrations of **1** and GF which induced 50% inhibition of cell growth when used in combination.

2.6. Effects of Combined Treatment of Targeted Agents and Penitrem A on Receptor Tyrosine Kinases (RTKs) and Their Downstream Effectors

Earlier studies showed that BK channels regulate Ca^{2+} signaling, which in turn mediates EGFR transactivation [54]. In this part, the effect of **1** and combined targeted therapy on total and active levels of RTKs, along with their downstream transducers, was evaluated using Western blot analysis. The concentrations of 0.5 μM of **1** and 60 nM LP, which resulted in synergistic antiproliferative activity in viability studies in vitro, were further applied to assess combination effects on downstream pathways. This combined LP and **1** treatment reduced the phosphorylated (active) levels of EGFR, HER2, and AKT, compared to individual treatments and vehicle-treated controls (Figure 14). However, total levels of EGFR and HER2 receptors did not remarkably change in response to treatment. In addition, there was an increase in the levels of the arrest protein p27 with combined treatment. Combination of **1**-LP mediated the inhibition of AKT signaling, leading to p27 stabilization and accumulation, which is a known inhibitor of CDK2. This resulted in G_1 arrest and subsequent growth inhibition in BT-474 cells [51,55].

Similarly, the synergistic combination of **1** at 0.5 μM with GF 125 nM caused a reduction in the levels of p-EGFR, p-HER2, p-AKT, and p-STAT3 (Figure 15). In addition, this combination upregulated the expression of the arrest protein p27.

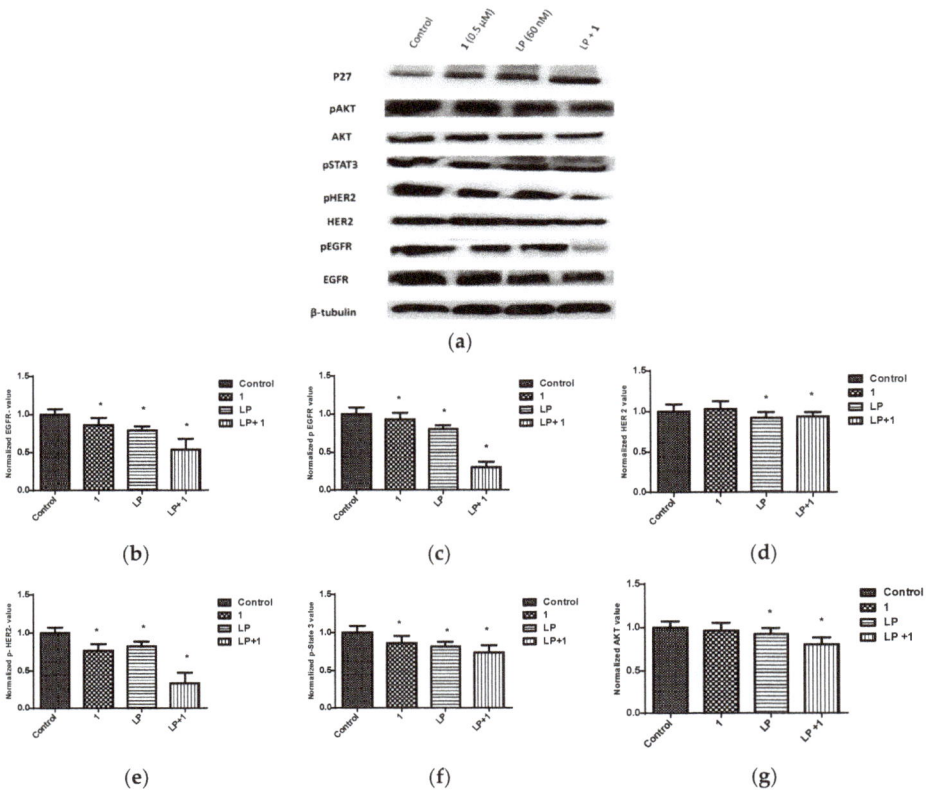

Figure 14. *Cont.*

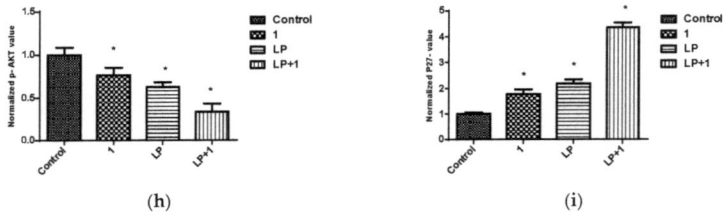

(h)

(i)

Figure 14. Effects of combined treatments of **1** and LP on RTKs and downstream effectors in BT-474 cells. (**a**) Western blots for BT-474 cells treated with vehicle control, **1** (0.5 µM), LP (60 nM), and **1**-LP combination. (**b–i**) Quantification analysis for Western blots for the effect of treatment on levels of EGFR, p-EGFR, HER2, p-HER2, p-STAT3, AKT, p-AKT, and p27, respectively. Vertical bars in the graphs indicate tubulin normalized protein value ± SEM. *: indicate significant differences ($p \leq 0.05$).

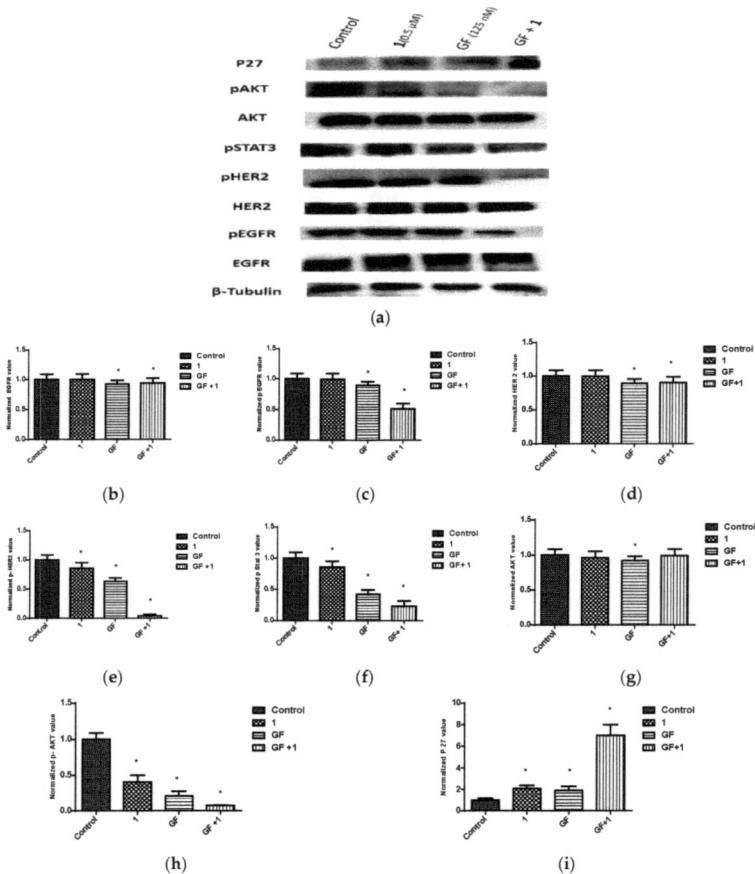

(a)

(b)

(c)

(d)

(e)

(f)

(g)

(h)

(i)

Figure 15. Effects of combined treatments of **1** and GF on RTK and downstream effectors in BT-474 cells. (**a**) Western blots for BT-474 cells treated with vehicle control, **1** (0.5 µM), GF (123 nM), and **1**-GF combination. (**b–i**) Quantification analysis for Western blots for the effect of treatment on levels of EGFR, p-EGFR, HER2, p-HER2, p-STAT3, AKT, p-AKT, and p27, respectively. Vertical bars in the graphs indicate tubulin normalized protein value ± SEM. *: indicate significant differences ($p \leq 0.05$).

3. Discussion

Three penitrem alkaloids, including **1** and **2**, previously isolated from the marine derived *P. commune* isolate GS20 [30,31], along with the semisynthetic **3**, were selected for testing their antiproliferative effects in BC cell lines in vitro [30,32]. This selection represents our previously reported pharmacophoric map of penitrems for anticancer activity [30,31]. Penitrem **1** represents the original natural parent of penitrems, while **2** is missing the important C-6 chloro substitution. The 25-*O*-methylated analog **3** is a semisynthetic product in which the free C-25 secondary alcohol pharmacophore was blocked by methylation.

Various K$^+$ channels are involved in the control of Ca^{2+} homeostasis, and hence, deregulation of their expression and/or activity will significantly affect cell proliferation [33]. Clearly, **1** is the most potent BK channel antagonist, and is associated with most remarkable antiproliferative effects among different BC cell lines, compared to other tested penitrems **2** and **3**. While growth inhibition of MDA-MB-231 and BT-474 with **1** was comparable, higher concentration of **1** was reported for the inhibition of SK-BR-3 cell growth. Although potentially toxic to non-tumorigenic cells, the IC$_{50}$ concentrations of **1** needed to suppress growth of both MCF-12A mammary epithelium and neuronal cells were multiple-fold higher than those required to inhibit BC cell growth (Table 1).

Penitrem **1** showed good fitting within the Ca^{2+} binding and Ca^{2+} bowel sites (Figures 2–4) in crystal structures 3MT5 and 3NAF, respectively, unlike **2** and **3**.

Among BC cell lines, highest expression levels were observed in the luminal B BT-474 cells. MDA-MB-231 and SK-BR-3 cells showed comparable levels KCNMA1. The lowest expression level of KCNMA1 was observed however in non-tumorigenic MCF-12A mammary epithelial cells (Figure 6).

TNF-α is a prototype cytokine which imparts a multitude of actions on many cell types, including neurons [40]. TNF-α was originally discovered as a serum factor causing necrosis of tumor cells [41]. TNF-α production in human macrophages is dependent on Ca^{2+}/calmodulin/calmodulin kinase II pathway [42]. Antagonism of BK channels has been shown to induce TNF-α production [40,43]. Taking into consideration that TNF-α is a major cytokine known to induce cancer cell death through sustained JNK-activation, it can be concluded that the antiproliferative activity of penitrems is associated, in part, with the upregulation of TNF-α and subsequent activation of apoptotic cell death [44,45].

Numerous reports showed the dependence of the cell cycle progression on the translocation of ions across the plasma membrane [13]. The antiproliferative activity of **1** was also shown to be mediated by the induction of G1 cell cycle arrest in BT-474 cells (Figure 11). The epidermal growth factor receptor (EGFR) and the epidermal growth factor receptor 2 (HER2) are ErbB family of receptor tyrosine kinases usually dysregulated in many BC subtypes [46]. Receptor tyrosine kinases (RTKs) are key regulators of cellular growth and proliferation, and dysregulation of RTKs promotes aggressive tumorigenesis pattern [47]. The EGFR/HER2 exert their biological effects through activating multiple downstream signaling cascades, including the PI3K/AKT, ERK1/2, and the transcription protein STAT [48]. Activated K$^+$ channel has been shown to increase proliferation rates in several BC cell lines [49,50]. The Ca^{2+}-activated K$^+$ currents have been observed in multiple BC cell lines, but their role was not clearly understood [34,39].

Small molecule tyrosine kinase inhibitors are increasingly recognized and utilized as treatments targeting specific signaling pathways associated with tumor development and progression. The small molecules lapatinib (LP) is a dual EGFR/HER2 inhibitor, while gefitinib (GF) is a selective EGFR inhibitor approved by FDA for oral control of HER-dependent malignancies, including BC. Despite the success of LP and GF as HER-targeting therapy, limitations to their clinical use involve significant hepatotoxicity and emergence of both primary and acquired resistance [3–5]. Many RTKs downstream effects are mediated by calcium signaling, which is regulated by BK channels [51].

Calculated CI of combined **1** with either LP or GF treatments indicated synergistic effects, with CI values less than 1. In addition, isobolograms showed synergistic interactions between **1** and both drugs as indicated by most data points located far below the line of additivity (Figures 12B and 13B). Combination synergy was also associated with reduced levels of p-EGFR, p-HER2, p-AKT, and p-STAT3, which are critical for cell survival and growth. Combinations also enhanced the level of the arrest protein p27. Reduced expression of p27 has also been shown to correlate with the ability of HER2-positive BC cells to escape HER2-targeting therapies, like trastuzumab [55]. The synergistic effect of these combinations is interesting, as it provides the opportunity to reduce the therapeutic doses of the HER family-targeted therapies LP and GF, which would ultimately reduce their potential toxicities and minimize the ability of HER2-dependent BC to develop resistance to targeted therapies.

Collectively, combined treatment of **1** and the anti-HER targeted therapies lapatinib and gefitinib showed remarkable synergistic antiproliferative activity associated with reduced target receptor activation, along with downstream effectors. This clearly highlights the role of BK channel as an appealing molecular target in BC.

4. Materials and Methods

4.1. Chemicals

All chemicals were purchased from Sigma-Aldrich (St. Louis, MO, USA), unless otherwise stated. Organic solvents were purchased from VWR (Suwanee, GA, USA), dried by standard procedures, packaged under nitrogen in Sure/Seal bottles, and stored over 4 Å molecular sieves. Unless otherwise indicated, cell culture reagents were obtained from Life Sciences (Carlsbad, CA, USA). Penitrems **1** and **2** were isolated from the marine-derived *Penicillium commune* isolate GS20, while penitrem **3** was semisynthetically prepared from **1** by direct methylation, as described earlier [30,31,49].

4.2. Docking Study

In silico binding studies were performed using Schrödinger molecular modeling software (v9, Schrodinger, New York, NY, USA), as previously described [56]. The calcium site domain of each of the BK channel crystal structure PDB codes 3NAF and 3MT5, resolution $\geq$3 Å, was prepared using the protein preparation wizard, applying PROPKA (Jensen Research Group, Denmark) to add hydrogens, assign bond orders, and optimize hydrogen bonding networks. The optimizing potentials for liquid stimulation (OPLS) force field (OPLS-2005, Schrodinger, New York, NY, USA) were used to ensure proper energy minimization (RMSD = 0.2 Å). The grid generation wizard was then applied on the resulting prepared BK channels structures to create the receptor energy grids as a cubic box centered on the co-crystallized ligand of each crystal structure, and extended for 10 Å in all directions. Penitrem analogs **1–3** were drawn using ChemDraw Professional 16 and imported as molfiles to the Maestro 9.3 panel interface (Maestro version 9.3, 2012, Schrodinger, New York, NY, USA). The LigPrep wizard using OPLS force field was chosen to prepare the 3D-dimensional structure of each analog, undergo geometric optimization, search for different conformers, and calculate partial atomic charges. Docking of penitrems was then conducted using the Glide 5.8 module (Glide, 2012, Schrodinger, New York, NY, USA) in extra-precision (XP) mode or in standard-precision mode (SP) [56].

4.3. Cell Viability Assay

Penitrems were dissolved in dimethyl sulfoxide (DMSO) to provide a final 25 mM stock solution. Proper media were used to prepare penitrems at their final tested doses for each assay. The DMSO vehicle control was prepared by adding the maximum volume of DMSO used in preparing tested penitrem to appropriate media type, in which the final DMSO concentration did not exceed 0.2% [30,31]. The effects of penitrems **1–3** on the proliferation and growth of multiple BC cell lines were evaluated using MTT assay. MDA-MB-231, BT-474, and SK-BR-3 BC cells (ATCC, Manassas, VA, USA) were initially seeded at 1×10^4 cells/well (6 replicates/group) in 96-well plates in RPMI-1640 media containing 10% FBS, and allowed to attach overnight. The next day, cells were washed with PBS, divided into different treatment groups, and then given various concentrations of tested penitrems in 0.5% FBS containing media. Viable cell number was calculated after 48 h using the MTT assay [30].

To evaluate the effect of penitrems on growth of the human neuronal Schwann CRL 2765 (ATCC, Manassas, VA, USA), cells were plated at a density of 1×10^4 cell per well (6-wells/group) in 96-well culture plates using the ATCC-recommended media (Dulbecco's Modified Eagle's Medium, DMEM) and allowed to adhere overnight for growth studies, after which each treatment was added, and cells were maintained for 48 h [49]. Similarly, the non-tumorigenic mammary epithelial MCF-12A cells were plated in DMEM/F12 supplemented with 5% horse serum, 0.5 mg/mL hydrocortisone, 20 ng/mL EGF, 100 U/mL penicillin, 0.1 mg/mL streptomycin, and 10 mg/mL insulin. Next day, cells were divided into different groups and fed serum-free DMEM media supplemented with experimental treatments or vehicle-treated control media. Viable cell number was determined using the MTT colorimetric assay. The absorbance was measured at λ_{570} nm on a BioTek Synergy2 microplate reader (BioTek, Winooski, VT, USA). The number of cells per well was calculated against a standard curve prepared at the start of the experiment. The IC_{50} value for each tested sample was calculated by non-linear regression of log concentration versus the % survival, implemented in GraphPad® PRISM version 5.0 (Graph Pad Software, Inc., La Jolla, CA, USA).

4.4. Western Blot Analysis

Western blot analysis was used to demonstrate the effects of penitrem treatments on BK channel and TNF-α expression level, and to study the combination effects of penitrem A with LP and GF. MDA-MB-231, BT-474, SK-BR-3, Schwann CRL-2765, and MCF-12A cells, were plated at a density of 1×10^6 cells/100 mm culture plates. MDA-MB-231, BT-474, and SK-BR-3 were plated in RPMI-1640 media supplemented with 10% FBS, Schwann cells CRL-2765 were plated in DMEM media supplemented with 10% FBS [11], and MCF-12A cells were plated in DMEM/F12 supplemented with ingredients described earlier. Cells were allowed to adhere overnight, then washed twice with PBS, and starved in control or treatment in serum-free medium containing 0.5% FBS for 48 h to synchronize the cells in G1 phase. Afterwards, cells were treated with tested penitrems in serum-free defined media for 24 h. At the end of treatment periods, cells were lysed in RIPA buffer (Qiagen Sciences Inc., Valencia, CA, USA). Protein concentration was determined by the BCA assay (Bio-Rad, Hercules, CA, USA). Equivalent amounts of protein (50 μg) were electrophoresed on SDS-polyacrylamide gels. The gels were then electroblotted onto PVDF membranes. These PVDF membranes were then blocked with 2% BSA in 10 mM Tris-HCl containing 50 mM NaCl and 0.1% Tween 20, pH 7.4 (TBST), and then incubated with specific primary antibodies overnight at 4 °C. The following dilutions were used—1:500 for BK channel antibody (Abbiotec LLC., San Diego, CA, USA) and 1:1000 for each of TNF-α, EGFR, pEGFR, STAT, pSTAT, HER2, pHER 2, AKT, pAKT, and p27 (Cell Signaling, Danvers, MA, USA). At the end of incubation period, membranes were washed 5 times with TBST, and then incubated with respective horseradish peroxide-conjugated secondary antibody in 2% BSA in TBST for 1 h at room temperature (rt), followed by rinsing with TBST for 5 times. Blots were then visualized and pictures taken using an enhanced chemiluminescence detection system according to the manufacturer's instructions (Chemi-Doc™ touch imaging system, Bio-Rad, Hercules, CA, USA). The visualization of β-tubulin was used to ensure equal sample loading in each lane. Western blot quantification analysis was done using Image Lab ™ Software (Bio-Rad, Hercules, CA, USA) for total protein normalization.

4.5. Immunocytochemistry

Immunocytochemical fluorescent studies were conducted by growing MDA-MB-231 and BT-474 cells mammary cancer cells on 8-chamber culture slides (Becton Dickinson and Company, Franklin Lakes, NJ, USA) at a density of 2×10^4 cells/chamber (2 replicates/group), and allowed to attach overnight in complete serum media supplemented with 10% FBS. Cells were then washed with Ca^{2+} and Mg^{2+}-free phosphate buffered saline (PBS), and incubated with vehicle control or treatment media containing 0.5% FBS for 48 h. At the end of experiment, cells were washed with pre-cooled PBS, and fixed with acetone, pre-cooled to -20 °C, for 2 min. Fixed cells were then washed with PBS, and blocked with 2% BSA in 10 mM Tris-HCl containing 50 mM NaCl and 0.1% Tween 20, pH 7.4 (TBST) for 1 h at rt. Cells were then incubated with specific primary antibody for BK channel KCNMA1 (1:300) overnight at 4 °C in 2% BSA-TBST. At the end of incubation time, cells were washed five times with pre-cooled PBS, followed by incubation with Alexa-Fluor 594-conjugated secondary antibody (1:5000) in 2% BSA-TBST for 1 h at rt. After final washing for five times with PBS, cells were embedded in Vectashield Mounting Medium with DAPI (Vector Laboratories Inc., Burlingame, CA, USA). Fluorescent images were obtained by using a confocal laser scanning microscope (Carl Zeiss Microimaging Inc., Thornwood, NY, USA). The red color indicates the positive fluorescence staining for BK channel KCNMA1, and the blue color represents MDA-MB-231 cell nuclei counter-stained with DAPI [30]. The experiment was repeated at least three times and multiple images for each chamber were captured. Magnification of each photomicrograph is 20×.

4.6. Analysis of Cell Cycle Progression by Flow Cytometry

To study **1** effect on cell cycle, BT-474 cells were plated, allowed to synchronize in G_1 phase, and then treated with **1** at concentrations of 0, 5, 10, 20, and 30 μM. At the end of the experiment, cells in different treatment groups were isolated with trypsin and then resuspended in ice-cold PBS, fixed with cold (−20 °C) 70% ethanol, and stored at 4 °C for 2 h. The cells were then rehydrated with ice-cold PBS and incubated with DNA staining buffer (sodium citrate 1 mg/mL, Triton X-100 3 μL/mL, PI 100 μg/mL, ribonuclease A 20 μg/mL) for 30 min at 4 °C in the dark. DNA content was analyzed using a FACScaliber flow cytometer (BD Biosciences, San Jose, CA, USA). For each sample, 10,000 events were recorded. A histogram was generated using Cell Quest software (BD Biosciences, San Jose, CA, USA). All experiments were repeated at least three times.

4.7. Statistical Analysis

The biochemical parameters were analyzed by repeated measures, one-way ANOVA followed by Tukey's test. Differences with $p < 0.05$ were considered statistically significant.

5. Conclusions

Findings from this study revealed remarkable expression of BK channels in multiple BC subtypes, and the potential pharmacological targeting for this channel. Penitrem **1** is a BK channel antagonist with promising antiproliferative activity in BC subtypes overexpressing BK channels. Combination of penitrem A with HER-targeting drugs, LP or GF, resulted in synergistic antiproliferative effects via STAT3 and p27 pathways (Figure 16), which could represent a novel strategy to improve cancer cell sensitivity to targeted regimens and reduce the emergence of resistance to these treatments.

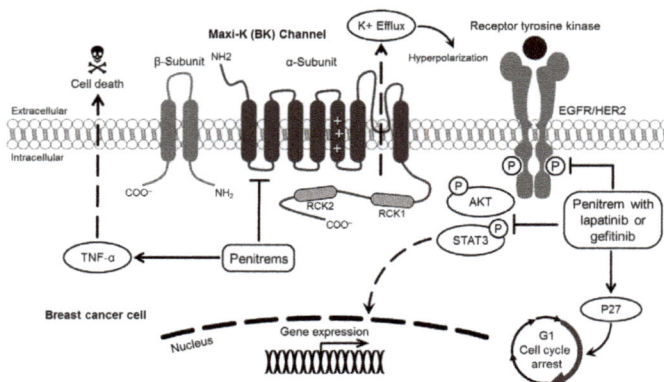

Figure 16. Summary of the pharmacological effects of the BK channel antagonist **1** and its combinations with the HER family-targeting drugs lapatinib and gefitinib in the human luminal B BT-474 BC cells.

Supplementary Materials: The following are available online at http://www.mdpi.com/1660-3397/16/5/157/s1, Figures S1–S3: Additional docking and binding modes of penitrems **2** and **3**.

Author Contributions: Conception and design: A.A.G., N.M.A., K.A.E.S.; Development of methodology: A.A.G., M.M., A.B.S.; Acquisition of data (provided animals, acquired and managed patients, provided facilities, etc.): A.A.G., A.B.S., M.M.; Analysis and interpretation of data (e.g., statistical analysis, biostatistics, computational analysis): A.A.G., N.M.A., M.M.; Writing, review, and/or revision of the manuscript: A.A.G., N.M.A., K.A.E.S.; Administrative, technical, or material support (i.e., reporting or organizing data, constructing databases): A.A.G.; Study supervision: K.A.E.S., N.M.A.

Acknowledgments: The financial support of the National Cancer Institute of the National Institutes of Health, Award Number R15CA167475 is acknowledged.

Conflicts of Interest: The authors declare no conflict of interest.

References

1. Siegel, R.L.; Miller, K.D.; Jemal, A. Cancer statistics, 2018. *CA Cancer J. Clin.* **2018**, *68*, 7–30. [CrossRef] [PubMed]

2. Polyak, K. Breast cancer: Origins and evolution. *J. Clin. Investig.* **2007**, *117*, 3155–3163. [CrossRef] [PubMed]

3. Goldhirsch, A.; Glick, J.H.; Gelber, R.D.; Coates, A.S.; Thurlimann, B.; Senn, H.J.; Panel, M. Meeting highlights: International expert consensus on the primary therapy of early breast cancer 2005. *Ann. Oncol.* **2005**, *16*, 1569–1583. [CrossRef] [PubMed]

4. Perou, C.M.; Sorlie, T.; Eisen, M.B.; van de Rijn, M.; Jeffrey, S.S.; Rees, C.A.; Pollack, J.R.; Ross, D.T.; Johnsen, H.; Akslen, L.A.; et al. Molecular portraits of human breast tumours. *Nature* **2000**, *406*, 747–752. [CrossRef] [PubMed]

5. Sorlie, T.; Tibshirani, R.; Parker, J.; Hastie, T.; Marron, J.S.; Nobel, A.; Deng, S.; Johnsen, H.; Pesich, R.; Geisler, S.; et al. Repeated observation of breast tumor subtypes in independent gene expression data sets. *Proc. Natl. Acad. Sci. USA* **2003**, *100*, 8418–8423. [CrossRef] [PubMed]

6. Tran, B.; Bedard, P.L. Luminal-B breast cancer and novel therapeutic targets. *Breast Cancer Res.* **2011**, *13*, 221. [CrossRef] [PubMed]

7. Hon, J.D.; Singh, B.; Sahin, A.; Du, G.; Wang, J.; Wang, V.Y.; Deng, F.M.; Zhang, D.Y.; Monaco, M.E.; Lee, P. Breast cancer molecular subtypes: From TNBC to QNBC. *Am. J. Cancer Res* **2016**, *6*, 1864–1872. [PubMed]

8. Toro, L.; Li, M.; Zhang, Z.; Singh, H.; Wu, Y.; Stefani, E. Maxi-K channel and cell signalling. *Pflugers Arch. Eur. J. Physiol.* **2014**, *466*, 875–886. [CrossRef] [PubMed]

9. Yang, H.; Zhang, G.; Cui, J. BK channels: Multiple sensors, one activation gate. *Front. Physiol.* **2015**, *6*, 29. [CrossRef] [PubMed]

10. Knaus, H.G.; Eberhart, A.; Glossmann, H.; Munujos, P.; Kaczorowski, G.J.; Garcia, M.L. Pharmacology and structure of high conductance calcium-activated potassium channels. *Cell. Signal.* **1994**, *6*, 861–870. [CrossRef]

11. Tanaka, Y.; Meera, P.; Song, M.; Knaus, H.G.; Toro, L. Molecular constituents of maxi KCa channels in human coronary smooth muscle: Predominant alpha + beta subunit complexes. *J. Physiol.* **1997**, *502 Pt 3*, 545–557. [CrossRef] [PubMed]

12. Vetri, F.; Choudhury, M.S.; Pelligrino, D.A.; Sundivakkam, P. BKCa channels as physiological regulators: A focused review. *J. Recept. Ligand Channel Res.* **2014**, *7*, 3–13. [CrossRef]

13. Ouadid-Ahidouch, H.; Roudbaraki, M.; Delcourt, P.; Ahidouch, A.; Joury, N.; Prevarskaya, N. Functional and molecular identification of intermediate-conductance Ca^{2+}-activated K^+ channels in breast cancer cells: Association with cell cycle progression. *Am. J. Physiol. Cell Physiol.* **2004**, *287*, C125–C134. [CrossRef] [PubMed]

14. Kunzelmann, K. Ion channels and cancer. *J. Membr. Biol.* **2005**, *205*, 159–173. [CrossRef] [PubMed]

15. Wonderlin, W.F.; Strobl, J.S. Potassium channels, proliferation and G1 progression. *J. Membr. Biol.* **1996**, *154*, 91–107. [CrossRef] [PubMed]

16. Pardo, L.A. Voltage-gated potassium channels in cell proliferation. *Physiology* **2004**, *19*, 285–292. [CrossRef] [PubMed]

17. Bloch, M.; Ousingsawat, J.; Simon, R.; Schraml, P.; Gasser, T.C.; Mihatsch, M.J.; Kunzelmann, K.; Bubendorf, L. KCNMA1 gene amplification promotes tumor cell proliferation in human prostate cancer. *Oncogene* **2007**, *26*, 2525–2534. [CrossRef] [PubMed]

18. Lang, F.; Foller, M.; Lang, K.S.; Lang, P.A.; Ritter, M.; Gulbins, E.; Vereninov, A.; Huber, S.M. Ion channels in cell proliferation and apoptotic cell death. *J. Membr. Biol.* **2005**, *205*, 147–157. [CrossRef] [PubMed]

19. Ousingsawat, J.; Spitzner, M.; Schreiber, R.; Kunzelmann, K. Upregulation of colonic ion channels in APC (Min/$^+$) mice. *Pflugers Arch. Eur. J. Physiol.* **2008**, *456*, 847–855. [CrossRef] [PubMed]

20. Spitzner, M.; Ousingsawat, J.; Scheidt, K.; Kunzelmann, K.; Schreiber, R. Voltage-gated K$^+$ channels support proliferation of colonic carcinoma cells. *FASEB J.* **2007**, *21*, 35–44. [CrossRef] [PubMed]

21. Kraft, R.; Krause, P.; Jung, S.; Basrai, D.; Liebmann, L.; Bolz, J.; Patt, S. BK channel openers inhibit migration of human glioma cells. *Pflugers Arch. Eur. J. Physiol.* **2003**, *446*, 248–255. [CrossRef] [PubMed]

22. Sontheimer, H. An unexpected role for ion channels in brain tumor metastasis. *Exp. Biol. Med.* **2008**, *233*, 779–791. [CrossRef] [PubMed]

23. Prevarskaya, N.; Skryma, R.; Shuba, Y. Ion channels and the hallmarks of cancer. *Trends Mol. Med.* **2010**, *16*, 107–121. [CrossRef] [PubMed]

24. Hanahan, D.; Weinberg, R.A. Hallmarks of cancer: The next generation. *Cell* **2011**, *144*, 646–674. [CrossRef] [PubMed]

25. Knaus, H.G.; McManus, O.B.; Lee, S.H.; Schmalhofer, W.A.; Garcia-Calvo, M.; Helms, L.M.; Sanchez, M.; Giangiacomo, K.; Reuben, J.P.; Smith, A.B., 3rd; et al. Tremorgenic indole alkaloids potently inhibit smooth muscle high-conductance calcium-activated potassium channels. *Biochemistry* **1994**, *33*, 5819–5828. [CrossRef] [PubMed]

26. Coiret, G.; Borowiec, A.S.; Mariot, P.; Ouadid-Ahidouch, H.; Matifat, F. The antiestrogen tamoxifen activates BK channels and stimulates proliferation of MCF-7 breast cancer cells. *Mol. Pharmacol.* **2007**, *71*, 843–8451. [CrossRef] [PubMed]

27. Valverde, M.A.; Rojas, P.; Amigo, J.; Cosmelli, D.; Orio, P.; Bahamonde, M.I.; Mann, G.E.; Vergara, C.; Latorre, R. Acute activation of Maxi-K channels (hSlo) by estradiol binding to the beta subunit. *Science* **1999**, *285*, 1929–1931. [CrossRef] [PubMed]

28. Steyn, P.S.; Vleggaar, R. Tremorgenic mycotoxins. *Prog. Chem. Org. Nat. Prod.* **1985**, *48*, 1–80.

29. Smith, M.M.; Warren, V.A.; Thomas, B.S.; Brochu, R.M.; Ertel, E.A.; Rohrer, S.; Schaeffer, J.; Schmatz, D.; Petuch, B.R.; Tang, Y.S.; et al. Nodulisporic acid opens insect glutamate-gated chloride channels: Identification of a new high affinity modulator. *Biochemistry* **2000**, *39*, 5543–5554. [CrossRef] [PubMed]

30. Sallam, A.A.; Ayoub, N.M.; Foudah, A.I.; Gissendanner, C.R.; Meyer, S.A.; El Sayed, K.A. Indole diterpene alkaloids as novel inhibitors of the Wnt/beta-catenin pathway in breast cancer cells. *Eur. J. Med. Chem.* **2013**, *70*, 594–606. [CrossRef] [PubMed]

31. Sallam, A.A.; Houssen, W.E.; Gissendanner, C.R.; Orabi, K.Y.; Foudah, A.I.; El Sayed, K.A. Bioguided discovery and pharmacophore modeling of the mycotoxic indole diterpene alkaloids penitrems as breast cancer proliferation, migration, and invasion inhibitors. *Med. Chem. Commun.* **2013**, *4*, 1360–1369. [CrossRef] [PubMed]

32. Wu, Y.; Yang, Y.; Ye, S.; Jiang, Y. Structure of the gating ring from the human large-conductance Ca^{2+}-gated K$^+$ channel. *Nature* **2010**, *466*, 393–397. [CrossRef] [PubMed]

33. Lallet-Daher, H.; Roudbaraki, M.; Bavencoffe, A.; Mariot, P.; Gackiere, F.; Bidaux, G.; Urbain, R.; Gosset, P.; Delcourt, P.; Fleurisse, L.; et al. Intermediate-conductance Ca^{2+}-activated K$^+$ channels (IKCa1) regulate human prostate cancer cell proliferation through a close control of calcium entry. *Oncogene* **2009**, *28*, 1792–1806. [CrossRef] [PubMed]

34. Yuan, P.; Leonetti, M.D.; Pico, A.R.; Hsiung, Y.; MacKinnon, R. Structure of the human BK channel Ca^{2+}-activation apparatus at 3.0 Å resolution. *Science* **2010**, *329*, 182–186. [CrossRef] [PubMed]

35. Hu, L.; Shi, J.; Ma, Z.; Krishnamoorthy, G.; Sieling, F.; Zhang, G.; Horrigan, F.T.; Cui, J. Participation of the S4 voltage sensor in the Mg^{2+}-dependent activation of large conductance (BK) K$^+$ channels. *Proc. Natl. Acad. Sci. USA* **2003**, *100*, 10488–10493. [CrossRef] [PubMed]

36. Bao, L.; Rapin, A.M.; Holmstrand, E.C.; Cox, D.H. Elimination of the BK(Ca) channel's high-affinity Ca^{2+} sensitivity. *J. Gen. Physiol.* **2002**, *120*, 173–189. [CrossRef] [PubMed]

37. Wonderlin, W.F.; Woodfork, K.A.; Strobl, J.S. Changes in membrane potential during the progression of MCF-7 human mammary tumor cells through the cell cycle. *J. Cell. Physiol.* **1995**, *165*, 177–185. [CrossRef] [PubMed]

38. Ouadid-Ahidouch, H.; Le Bourhis, X.; Roudbaraki, M.; Toillon, R.A.; Delcourt, P.; Prevarskaya, N. Changes in the K$^+$ current-density of MCF-7 cells during progression through the cell cycle: Possible involvement of a h-ether.a-gogo K$^+$ channel. *Recept. Channels* **2001**, *7*, 345–356. [PubMed]

39. Wegman, E.A.; Young, J.A.; Cook, D.I. A 23-pS Ca^{2+}-activated K$^+$ channel in MCF-7 human breast carcinoma cells: An apparent correlation of channel incidence with the rate of cell proliferation. *J. Physiol.* **1991**, *417*, 562–570. [CrossRef]

40. Fiers, W. Tumor necrosis factor. Characterization at the molecular, cellular and in vivo level. *FEBS Lett.* **1991**, *285*, 199–212. [CrossRef]

41. Carswell, E.A.; Old, L.J.; Kassel, R.L.; Green, S.; Fiore, N.; Williamson, B. An endotoxin-induced serum factor that causes necrosis of tumors. *Proc. Natl. Acad. Sci. USA* **1975**, *72*, 3666–3670. [CrossRef] [PubMed]

42. Chen, R.; Ji, G.; Wang, L.; Ren, H.; Xi, L. Activation of ERK1/2 and TNF-alpha production are regulated by calcium/calmodulin signaling pathway during *Penicillium marneffei* infection within human macrophages. *Microb. Pathog.* **2016**, *93*, 95–99. [CrossRef] [PubMed]

43. Enomoto, K.; Furuya, K.; Maeno, T.; Edwards, C.; Oka, T. Oscillating activity of a calcium-activated K$^+$ channel in normal and cancerous mammary cells in culture. *J. Membr. Biol.* **1991**, *119*, 133–139. [CrossRef] [PubMed]

44. Wang, X.; Lin, Y. Tumor necrosis factor and cancer, buddies or foes? *Acta Pharmacol. Sin.* **2008**, *29*, 1275–1288. [CrossRef] [PubMed]

45. Balkwill, F. Tumour necrosis factor and cancer. *Nat. Rev. Cancer* **2009**, *9*, 361–371. [CrossRef] [PubMed]

46. Foley, J.; Nickerson, N.K.; Nam, S.; Allen, K.T.; Gilmore, J.L.; Nephew, K.P.; Riese, D.J. EGFR signaling in breast cancer: Bad to the bone. In *Seminars in Cell & Developmental Biology*; Academic Press: Cambridge, MA, USA, 2010; Volume 21, pp. 951–960.

47. Aaronson, S.A. Growth factors and cancer. *Science* **1991**, *254*, 1146–1153. [CrossRef] [PubMed]

48. Yarden, Y.; Sliwkowski, M.X. Untangling the ErbB signalling network. *Nat. Rev. Mol. Cell Biol.* **2001**, *2*, 127–137. [CrossRef] [PubMed]

49. Goda, A.A.; Naguib, K.M.; Mohamed, M.M.; Amra, H.A.; Nada, S.A.; Abdel-Ghaffar, A.B.; Gissendanner, C.R.; El Sayed, K.A. Astaxanthin and docosahexaenoic acid reverse the toxicity of the maxi-K (BK) channel antagonist mycotoxin penitrem A. *Mar. Drugs* **2016**, *14*, 208. [CrossRef] [PubMed]

50. Sings, H.; Singh, S. Tremorgenic and nontremorgenic 2,3-fused indole diterpenoids. *Alkaloids Chem. Biol.* **2003**, *60*, 51–163. [PubMed]

51. Chu, I.M.; Hengst, L.; Slingerland, J.M. The CDK inhibitor p27 in human cancer: Prognostic potential and relevance to anticancer therapy. *Nat. Rev. Cancer* **2008**, *8*, 253–267. [CrossRef] [PubMed]

52. Chou, T.C. Drug combination studies and their synergy quantification using the Chou-Talalay method. *Cancer Res.* **2010**, *70*, 440–446. [CrossRef] [PubMed]

53. Tallarida, R.J. An overview of drug combination analysis with isobolograms. *J. Pharmacol. Exp. Ther.* **2006**, *319*, 1–7. [CrossRef] [PubMed]

54. Oda, K.; Matsuoka, Y.; Funahashi, A.; Kitano, H. A comprehensive pathway map of epidermal growth factor receptor signaling. *Mol. Syst. Biol.* **2005**, *1*. [CrossRef] [PubMed]

55. Shattuck, D.L.; Miller, J.K.; Carraway, K.L., 3rd; Sweeney, C. Met receptor contributes to trastuzumab resistance of Her2-overexpressing breast cancer cells. *Cancer Res.* **2008**, *68*, 1471–1477. [CrossRef] [PubMed]

56. Mohyeldin, M.M.; Busnena, B.A.; Akl, M.R.; Dragoi, A.M.; Cardelli, J.A.; El Sayed, K.A. Novel c-Met inhibitory olive secoiridoid semisynthetic analogs for the control of invasive breast cancer. *Eur. J. Med. Chem.* **2016**, *118*, 299–315. [CrossRef] [PubMed]

marine drugs

MDPI

Article

Raistrickiones A—E from a Highly Productive Strain of *Penicillium raistrickii* Generated through Thermo Change

De-Sheng Liu [1], Xian-Guo Rong [1], Hui-Hui Kang [1], Li-Ying Ma [1], Mark T. Hamann [2] and Wei-Zhong Liu [1,*]

[1] College of Pharmacy, Binzhou Medical University, Yantai 264003, China; desheng_liu@sina.com (D.-S.L.); binyirongxianguo@163.com (X.-G.R.); kanghuihui_1993@126.com (H.-H.K.); maliyingbz@163.com (L.-Y.M.)
[2] Department of Drug Discovery and Biomedical Sciences, Medical University of South Carolina, Charleston, SC 29425, USA; hamannm@musc.edu
*   Correspondence: lwz1963@163.com; Tel.: +86-535-691-3205

Received: 25 May 2018; Accepted: 15 June 2018; Published: 18 June 2018

Abstract: Three new diastereomers of polyketides (PKs), raistrickiones A—C (**1–3**), together with two new analogues, raistrickiones D and E (**4 and 5**), were isolated from a highly productive strain of *Penicillium raistrickii*, which was subjected to an experimental thermo-change strategy to tap its potential of producing new secondary metabolites. Metabolites **1** and **2** existed in a diastereomeric mixture in the crystal packing according to the X-ray data, and were laboriously separated by semi-preparative HPLC on a chiral column. The structures of **1–5** were determined on the basis of the detailed analyses of the spectroscopic data (UV, IR, HRESIMS, 1D, and 2D NMR), single-crystal X-ray diffractions, and comparison of the experimental and calculated electronic circular dichroism spectra. Compounds **1–5** represented the first case of 3,5-dihydroxy-4-methylbenzoyl derivatives of natural products. Compounds **1–5** exhibited moderate radical scavenging activities against 1,1-diphenyl-2-picrylhydrazyl radical 2,2-diphenyl-1-(2,4,6-trinitrophenyl) hydrazyl (DPPH).

Keywords: *Penicillium raistrickii*; polyketides; diastereomers; thermo-change strategy

1. Introduction

A plethora of publications have already delineated fungal strains from unique environments, including saline soil near the ocean, marine biospheres, and other special niches. Fungi from these locations have been shown to be an excellent biosynthetic source of chemical diversity and secondary metabolites (SMs) for pharmaceutical applications [1–5]. With the significant advancements in genomics and metagenomics, it has been clearly shown that the capacity for fungal strains to produce small molecules is determined by various biosynthetic gene clusters (BGCs) of their genome [6,7]. However, under routine or constant laboratory culture conditions, large amounts of BGCs remained "silent" and unexpressed [8], leading to the limited categories or numbers of SMs being produced. More recently, it was progressively elucidated that the silent BGCs could be successfully activated by manipulation of the culture conditions, such as cultivation-based and molecular approaches [9–12]. Among cultivation-based approaches, thermo change is one of the effective strategies for triggering silent biosynthetic expression systems to enlarge the numbers of fungi-derived natural products, with only a few cases reported [13–15]. Strain JH-18 of *P. raistrickii*, which was isolated from marine saline soil of the coast of Bohai Bay in China, was investigated with only a few reports [16–19]. Its routine laboratory fermentation enabled the isolation of series of novel natural products including spiroketals, isocoumarins, α-pyrone, and dihydropyran derivatives [16–18], and some of them possess an unusual chemical skeleton and exhibit cytotoxic activity [20].

In view of the foregoing findings, a series of OSMAC (one strain–many compounds) [21] protocols were performed to investigate the possibility of unlocking the silent genes of this prolific genus to generate more new SMs. During the investigation, it was discovered that the thermo-change approach worked very well in evoking the synthetic expression systems. Five new polyketide (PKs), named raistrickiones A−E (**1–5**), which showed structural differences with those reported previously, were isolated in response to the fermentation temperature setting at 15 °C instead of 28 °C, with other conditions unchanged. In the current work, the isolation, purification, structure, elucidation, and biological evaluation of compounds **1–5** (Figure 1) were carried out.

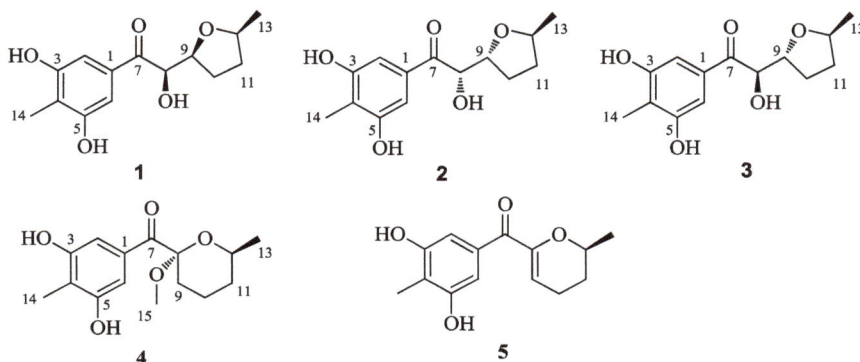

Figure 1. Structures of compounds **1–5**.

2. Results

Raistrickione A (**1**) was separated as colorless plates in MeOH, and its molecular formula was determined as $C_{14}H_{18}O_5$, with six indices of hydrogen deficiency, by the deprotonated molecular ion peak at *m/z* 265.1083 [M − H]$^-$ (calculated for $C_{14}H_{17}O_5$, 265.1081). The IR spectrum exhibited absorption bands for the presence of hydroxy (3259 cm^{-1}, broad), keto carbonylic (1677 cm^{-1}), and aromatic (1592 and 1421 cm^{-1}) functionalities. The ^{1}H NMR data (Table 1) in DMSO-d_6 indicated two overlapped aromatic protons (δ_H 6.90, 2H, s), three hydroxy groups (two phenol and one alcoholic at δ_H 9.51, 2H, s; 4.84, 1H, d, *J* = 7.3 Hz, respectively), three oxygenated methines (δ_H 3.77, 1H, m; 4.13, 1H, m; 4.74, 1H, dd, *J* = 7.3, 3.5 Hz), two methylenes (δ_H 1.34, 1H, m; 1.84, 3H, m), and two methyls (δ_H 1.06, 3H, d, *J* = 6.0 Hz; 1.99, 3H, s). Analyses of the ^{13}C NMR (DMSO-d_6, Table 1) and DEPT data (Figure S17) demonstrated 14 carbon resonances, including two methyls, two methylenes, three oxygenated methines, one keto carbonyl, and six aromatic carbons (two tertiary and two oxygenated quaternary overlapped, respectively). The above information suggested a symmetrically tetrasubstituted phenyl ring in **1**. The upfield chemical shift (δ_C 8.9) of the arylmethyl group implied that it was sandwiched between two phenolic hydroxyls in the phenyl ring [16,17]. Furthermore, the upfield chemical shift (δ_C 199.2) of keto carbonyl suggested it to be conjugated with the phenyl ring, and located at the *para* position of the arylmethyl, which was proven by the non-chelated phenolic hydroxyl signals at δ_H 6.89. The presence of a 3,5-dihydroxyl-4-methylbenzoyl moiety was further disclosed by the associated HMBC correlations (Figure 2). The proton–proton correlation spectroscopy (^{1}H-^{1}H COSY, Figure 2) revealed a carbon chain from C-8 to C-13 (Figure 2), CHCHCH$_2$CH$_2$CHCH$_3$, corresponding to all of the other carbons. Considering the ^{1}H-^{1}H COSY correlation of the alcoholic hydroxyl (δ_H 4.84) with H-8, C-8 connected with the hydroxyl, and the other two oxymethines at δ_C 74.9 (C-9) and 75.2 (C-12) should be linked through an oxygen atom to form a substituted tetrahydrofuran ring to meet the remaining index of hydrogen deficiency. The two moieties, a 3,5-dihydroxyl-4-methylbenzoyl unit and a disubstituted tetrahydrofuran ring, were linked together by the HMBC cross-peaks from the alcoholic hydroxyl to C-7, C-8, and C-9, and from H-8 (δ_H 4.74) to C-7 and C-9, respectively. Based on these results, the 2D structure of **1** was established and in agreement with all of the HSQC and HMBC data. In the NOESY spectrum of **1** (Figure 3), the correlations between H-9 and H-12 indicated that the two hydrogens occupied the same face. Finally, the absolute configuration of **1** was fully established as 8*R*, 9*S*, and 12*S* by the X-ray crystallography results, in which the tetrahydrofuran ring has two conformations in view of the disorder from C-10 to C-13 (Figure 4). The absolute configuration of **1** was further confirmed by the experimental and calculated electronic circular dichroism (ECD) data (Figure 5).

Table 1. NMR Spectroscopic Data (^{1}H 400 MHz and ^{13}C 100 MHz) of **1–3** (DMSO-d_6).

Position	1		2		3	
	δ_C	δ_H (*J* in Hz)	δ_C	δ_H (*J* in Hz)	δ_C	δ_H (*J* in Hz)
1	133.0, C		132.9, C		133.5, C	
2, 6	106.2, CH	6.90, s	106.2, CH	6.89, s	106.2, CH	6.94, s
3, 5	156.2, C		156.2, C		156.3, C	
4	116.6, C		116.6, C		116.7, C	
7	199.0, C		199.2, C		199.4, C	
8	74.9, CH	4.74, dd (7.3, 3.5)	75.4, CH	4.69, dd (7.3, 3.3)	74.8, CH	4.73, t (5.8)
9	80.1, CH	4.13, m	79.5, CH	4.27, td (7.0, 3.3)	79.3, CH	4.20, q (5.8)
10	27.3, CH$_2$	1.84, ^{a} m	27.8, CH$_2$	1.93, ^{a} m	26.5, CH$_2$	1.84, m; 1.77, m
11	32.5, CH$_2$	1.84, ^{a} m; 1.34, m	33.4, CH$_2$	1.93, ^{a} m; 1.31, m	33.1, CH$_2$	1.99, m; 1.34, m
12	75.2, CH	3.77, m	75.4, CH	3.99, m	75.0, CH	4.04, m
13	20.6, CH$_3$	1.06, d (6.0)	21.0, CH$_3$	1.02, d (6.1)	21.0, CH$_3$	1.04, d (6.0)
14	8.9, CH$_3$	1.99, s	8.9, CH$_3$	1.99, s	8.9, CH$_3$	1.99, s
15						
OH-3, 5		9.51, s		9.52, s		9.49, s
OH-8		4.84, d (7.3)		4.96, d (7.3)		5.31, d (5.8)

^{a} Overlapping signals.

Raistrickione B (**2**) was also obtained as colorless plates. The negative HRESIMS data indicated that **2** has the same molecular formula of $C_{14}H_{18}O_5$ as that of **1**. Its ^{1}H and ^{13}C NMR spectra (Table 1) highly resembled those of **1** with slight differences in the substituted tetrahydrofuran moiety, which further proved that **2** is a diasteremomer of **1** by 2D NMR, data as described above. The electronic circular dichroism (ECD) data (Figure S11) of **2** exhibited almost opposite-sign bands all through the spectrum in comparison with that of **1** (Figure S4). The cotton effects in the ECD spectra suggested the two compounds had an opposite configuration at C-8, and hence an 8*S* absolute configuration in **2**. The configuration of the chiral center of C-9 was assigned by comparison of the vicinal proton–proton coupling constants between H-8 and H-9 in compounds **1** and **2**, which exhibited almost an identical magnitude ($^3J_{HH}$ = 3.5 and 3.3 Hz for **1** and **2**, respectively). According to the Karplus equation [22], the dihedral angle between H-8 and H-9 in compounds **1** and **2** should possess the same geometric behavior in their relative spatial arrangement, which consequentially led to the identification of the 9*R* absolute configuration in **2**. Based on the X-ray crystallography (Figure 4) data, the absolute configuration of **2** was fully established as 8*S*, 9*R*, and 12*S*. This was coincident with the optical rotation values (+47.5 and −48.6 for **1** and **2** in MeOH, respectively), which intensified the validity of the absolute configuration of **2**.

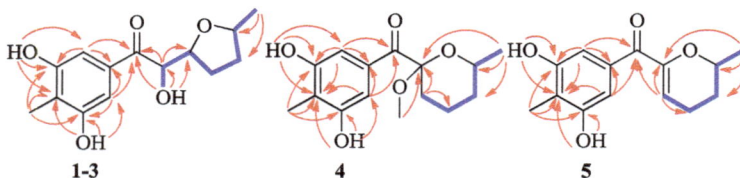

Figure 2. Key HMBC (red →) and ^{1}H–^{1}H COSY (blue —) correlations of **1–5**.

Raistrickione C (**3**) was afforded as colorless powder. The HRESIMS data assigned the same molecular formula as those of **1** and **2**. The ^{1}H and ^{13}C NMR spectra of **3** (Table 1) were closely similar to those of **2**. All of the information illustrated that it was another diastereoisomer that was different from **1** and **2**. The identical ECD data of **1** and **3** were reminiscent of the same configurational behavior, and hence an 8*R* absolute configuration in **3**. The t multiplicity (dd in **2**) of H-8 and the larger *J* value ($^3J_{HH}$ = 5.8 Hz) between H-8 and H-9 ($^3J_{HH}$ = 3.3 Hz in **2**) in **3** were indicative of the absolute configuration of C-9 to be *R*. In the NOESY spectrum (Figure S31), no cross-peak was observed between H-9 and H-12, so the H-9 and H-12 should be in *trans* orientations. The configuration of C-12 should be the same as in **2**, which was supported by the high similarity of ^{1}H and ^{13}C NMR data (Table 1) among C-11, C-12, and C-13 within compounds **2** and **3**. Consequently, the absolute configuration of **3** was established as 8*R*, 9*R*, and 12*S*.

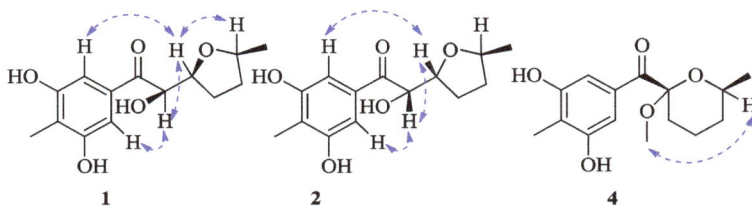

Figure 3. Key NOESY correlations (dashed blue arrows) of **1**, **2** and **4**.

Raistrickione D (**4**) was isolated as colorless needles in MeOH. The molecular formula of **4** was determined on the basis of the NMR data and the HRESIMS results, which showed a deprotonated molecular ion peak [M − H]$^−$ at *m/z* 279.1236 (calculated for C$_{15}$H$_{19}$O$_5$, 279.1238) with 14 *amu* more than those of compounds **1–3**, accounting for six indices of hydrogen deficiency. In the IR spectrum of **4**, absorptions at 3337 (broad) cm^{-1}, 1668 cm^{-1}, and 1588 cm^{-1}, were assigned to hydroxy, carbonylic, and aromatic functionalities, respectively. The UV spectrum (Figure S34) exhibited similar absorptions with those of **1–3**, indicating that **4** was an analogue of those compounds. The ^{1}H and ^{13}C NMR spectra (Table 2) presented a keto carbonyl (δ_C 196.5), two overlapped oxygenated aromatic quaternary carbons (δ_C 156.8), two aromatic quaternary carbons (δ_C 133.9, 117.6), two overlapped aromatic methines (δ_C 109.2; δ_H 7.36, 2H, s), an aromatic methyl (δ_C 9.0; δ_H 2.13, 3H, s), and two phenol hydroxyls (δ_H 8.43, 2H, s), which suggested there is a 3,5-dihydroxy-4-methylbenzoyl moiety in **4** as in **1**. In addition, one dioxygenated quaternary carbon, one methoxyl, one methyl, three methylenes, and one methine were observed in the NMR spectra of **4**. The COSY system (Figure 2) presented a proton spin system from H-9 to H-13, which anchored at the keto carbonyl (C-7) through the dioxygenated carbon (C-8) according to the HMBC correlations (Figure 2) from H-9 to C-7 and C-8. The HMBC correlation from the methoxyl (H-15) to the dioxygenated carbon (C-8) established its placement. The remaining degree of unsaturation completed the tetrahydropyran ring. The NOESY spectrum (Figure 3) of **4** exhibited cross-peaks between H-15 and H-12, indicating that the methoxyl and H-12 were on the same face of the tetrahydropyran ring, which determined the relative configuration of C-8 and C-12. The absolute configuration of **4** was solved by comparison of its experimental ECD spectrum with the predicted one by time-dependent density functional theory (TDDFT) calculation at the B3LYP/6-311G (d, p) level. As a result, the calculated ECD curve of (8*S*, 12*S*)-**4** (Figure 5) was in line with the experimental one. Therefore, the absolute configuration of **4** was elucidated as 8*S*, 12*S*.

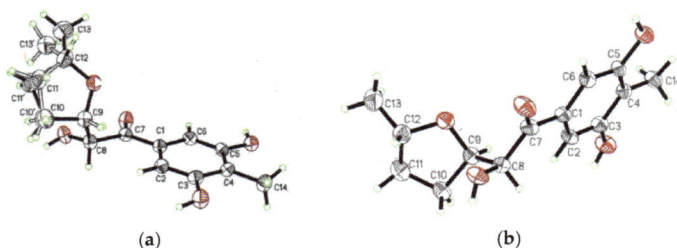

(a)   (b)

Figure 4. X-ray ORTEP drawings of **1** (a) and **2** (b).

Table 2. NMR Spectroscopic Data (^{1}H 400 MHz and ^{13}C 100 MHz) of **4** and **5**.

Position	4 ^{a}		5 ^{a}	
	δ_C	δ_H (*J* in Hz)	δ_C	δ_H (*J* in Hz)
1	133.9, C		136.3, C	
2, 6	109.2, CH	7.36, s	108.7, CH	6.94, s
3, 5	156.8, C		156.8, C	
4	117.6, C		116.8, C	
7	196.5, C		190.1, C	
8	102.6, C		152.7, C	
9	32.3, CH_2	1.80, m; 1.65, ^{b} m	112.2, CH	5.71, t (3.8)
10	19.7, CH_2	1.91, m; 1.65, ^{b} m	21.6, CH_2	2.30, m; 2.20, m
11	32.7, CH_2	1.65, ^{b} m; 1.40, m	29.3, CH_2	1.94, m; 1.55, m
12	68.0, CH	3.88, m	73.0, CH	4.04, m
13	22.0, CH_3	1.22, d (6.3)	21.2, CH_3	1.32, d (6.2)
14	9.0, CH_3	2.13, s	9.0, CH_3	2.15, s
15	50.5, CH_3	3.17, s		
OH-3, 5		8.43, s		8.45, s
OH-8				

^{a} NMR spectra obtained in acetone-d_6. ^{b} Overlapping signals.

Raistrickione E (**5**) was afforded as colorless powder. Its molecular formula $C_{14}H_{16}O_4$ was given based on the HREISMS data, which exhibited a deprotonated molecular ion peak [M − H]$^-$ at *m/z* 247.0976 (calculated for $C_{14}H_{15}O_4$, 247.0976), with seven indices of hydrogen deficiency. The ^{1}H and ^{13}C NMR data (Table 2) in acetone-d_6 were similar to those of **4**, except for the absence of a methoxyl, a dioxygenated quarternary carbon, and a methylene which existed in **4**, and the appearance of an oxygenated trisubstituted double bond (δ_C 152.7, 112.2; δ_H 5.71, 1H, t, *J* = 3.8 Hz) in **5**. This implied that there was a double bond between C-8 and C-9, which was further confirmed by the COSY spin system from H-9 to H-13, along with the HMBC correlations of H-9 with C-7, C-8, and C-10, and H-13 with C-8, C-11, and C-12 (Figure 2). Therefore, the two-dimensional (2D) structure of **5** was established. Similarly, the stereochemistry of **5** was assigned as 12*S* by comparing its experimental and theoretical ECD data (Figure 5).

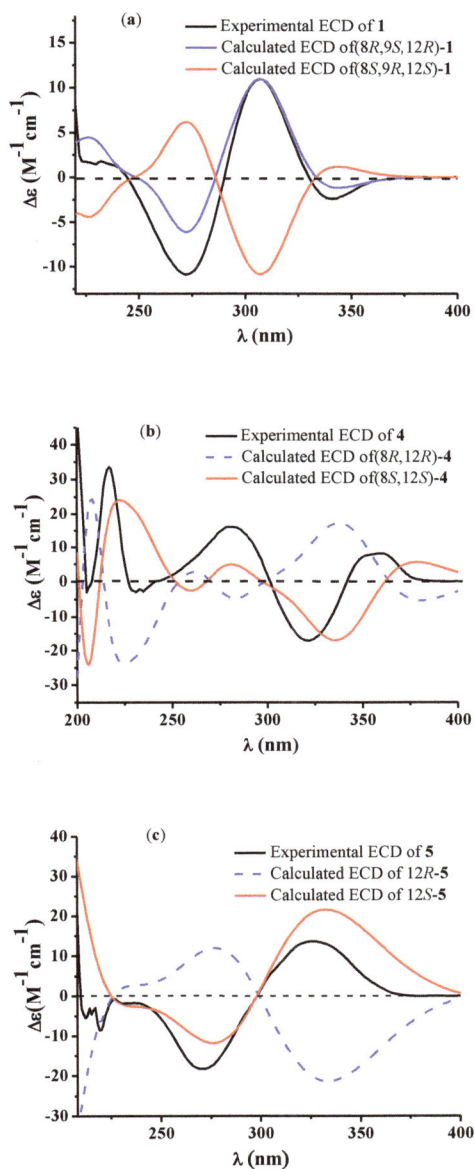

Figure 5. Experimental and calculated ECD of **1** (**a**), **4** (**b**) and **5** (**c**).

Compounds **1**–**5** exhibited moderate radical scavenging activities against DPPH with IC_{50} values of 32 ± 2.5 μM, 38 ± 1.9 μM, 40 ± 3.6 μM, 49 ± 2.1 μM, and 42 ± 1.2 μM, respectively; ascorbic acid was applied as a positive control (IC_{50}: 17 ± 1.7 μM). They were further evaluated for their cytotoxic effect against human leukemia (HL60) cell lines by the MTT method, and all of them were inactive (>20 μM).

3. Discussion

Intriguingly, a mixture of **1** and **2** was superficially obtained as a pure chemical at first and displayed characteristics of a dimer of compound **3** in NMR spectroscopic data, which was contradicted by the mass spectrometric results. Those observations suggested that it was a partial racemate in nature. This assumption was subsequently verified by the single-crystal X-ray diffraction analysis (Figure 4). Then, a preparative chiral HPLC was applied to separate **1** and **2** in an extremely time-consuming and repeated purification process, with a ratio of ca. 1:1 (Figure 6). The former peak centered at 17.72 min (retention time) was identified as **2**, and the latter one centered at 18.53 min was identified as **1**. Detailed comparison of the spectroscopic data revealed that the ^{1}H and ^{13}C NMR spectra of the isolated diastereoisomers assembled those of their precursor mixture (Figure S5, Figure S12, and Figure S15, ^{1}H spectra for **1**, **2**, and the precursor mixture, respectively; Figure S6, Figure S13, and Figure S16, ^{13}C spectra for **1**, **2**, and the precursor mixture, respectively). Unlike enantiomers, most diastereomers are easy to be separated by reverse phase HPLC or normal preparative thin layer chromatography (TLC) [23–28], due to their different physical properties. Only a few diastereomers need to use chiral HPLC to isolate [29,30].

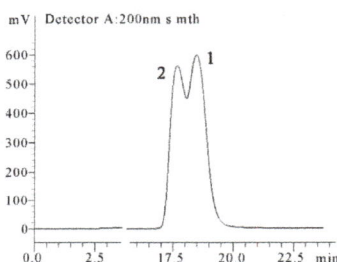

Figure 6. Separation of **1** and **2** on a chiral HPLC column.

4. Materials and Methods

4.1. General Experimental Procedures

Melting points were determined with an XRC-1 micro-melting point apparatus (Sichuan University Scientific Instrument Factory, Chengdu, China) and were uncorrected. Optical rotations were measured on an Autopol V Plus digital polarimeter (Rudolph Research Analytical, Hackettstown, NJ, USA). UV spectra were obtained on a TU-1091 spectrophotometer (Beijing Purkinje General Instrument Co., Beijing, China). ECD spectra were recorded with a Chirascan spectropolarimeter (Applied Photophysics, Leatherhead, United Kingdom). IR spectra were carried out on a Nicolet 6700 spectrophotometer (Thermo Scientific, Waltham, MA, USA) by an attenuated total reflectance (ATR) approach. NMR data were obtained at 400 MHz and 100 MHz for ^{1}H and ^{13}C, respectively, on an Avance 400 (Bruker, Billerica, MS, USA) with TMS as the internal standard. Crystal structure determination was performed on a Bruker Smart 1000 CCD X-ray diffractometer (Bruker Biospin Group, Karlstuhe, Germany). HRESIMS was acquired on a 1200RRLC-6520 Accurate-Mass Q-TOF LC/MS mass spectrometer (Agilent, Santa Clara, CA, USA). Semi-preparative HPLC was accomplished on a LC-6AD Liquid Chromatography (Shimadzu, Kyoto, Japan) with an SPD-20A Detector by an ODS column (HyperClone 5 μm ODS (C$_{18}$) 120 Å, 250 × 10 mm, Phenomenex, 4 mL/min.) Chiral HPLC was carried out on a column [ChiralPAK IC, 5 μm cellulose tri(3,5-dichlorophenyl carbamate), 250 × 10 mm, Daicel Chiral Technologies Co. LTD. (Shaihai, China), 4 mL/min]. Sephadex LH-20 (Ge Healthcare Bio-Sciences AB, Uppsala, Sweden), silica gel (200−300 mesh, Qingdao Marine Chemical Inc., Qingdao, China), and reversed-phase C$_{18}$ silica gel (Pharmacia Fine Chemical Co., Ltd., Uppsala, Sweden) were used for open column chromatography.

4.2. Fungal Material

Strain JH-18 of *P. raistrickii* (Genbank accession No. HQ717799), was isolated from the marine saline soil as previously reported [18]. The strain is kept at College of Pharmacy, Binzhou Medical University.

4.3. Fermentation, Extraction, and Isolation

The fermentation and extraction procedures were almost the same as described in a previous article [18], except for the fermentation temperature setting at 15 °C instead of 28 °C. The whole culture broth (40 L) afforded 23 g of crude extract. The extract was subjected to a silica gel column, eluting with different solvents of increasing polarity from petroleum ether, chloroform to MeOH to yield eight fractions (Fr.s 1–8) based on TLC analysis. Fr. 4 (12 g) was passed through a reversed-phase column eluting with MeOH-water gradient (from 20:80 to 100:0) to afford nine subfractions (Fr.s 4.1–4.9). Fr. 4.3 (2.8 g) was further separated on a Sephadex LH-20 column eluting with MeOH to afford five subfractions (Fr.s 4.3.1–4.3.5). Fr. 4.3.2 (98.1 mg) was purified by semipreparative HPLC on an ODS column eluting with MeOH-0.2% trifluoroacetic acid (TFA) aqueous solution (*v/v*) (4:6, 4 mL/min) to yield the diastereomeric mixture (29.2 mg, t_R 12.5 min) of **1** and **2**. Then, the mixture was separated on semi-preparative HPLC using the chiral column (hexane-isopropanol, 9:1; 4 mL/min) to yield compounds **1** (12.3 mg, t_R 18.53 min) and **2** (10.0 mg, t_R 17.72 min). Fr. 4.3.4 (68.0 mg) was reloaded on semi-preparative HPLC using an ODS column eluting with MeOH-0.2% TFA aqueous solution (5:5; 4 mL/min) to afford compound **3** (12.0 mg, t_R 23.5 min). Fr. 4.4 (1.1 g) was passed through a Sephadex LH-20 column to yield subfractions (Fr.s 4.4.1–4.4.8) eluting with MeOH. Purification of Fr. 4.4.5 (30.1 mg) by semi-preparative HPLC on an ODS column (MeOH-0.2% TFA aqueous solution; 4 mL/min) afforded compound **4** (9.0 mg, t_R 13.6 min). Fr. 4.6 (0.9 g) was further chromatographed on CC of Sephadex LH-20, developed by MeOH and then purified by semipreparative HPLC on an ODS column (MeOH-0.2% TFA aqueous solution, 5:5; 4 mL/min) to yield compound **5** (11.0 mg, t_R 9.1 min).

Raistrickione A (**1**): colorless plates (MeOH); mp 198–199 °C; $[\alpha]_D^{25}$ +47.5 (*c* 0.064, MeOH); UV (MeOH) λ_{max} (log ε) 220 (4.20), 279 (3.86) nm; IR (ATR) v_{max} 3259 (broad), 1677, 1592, 1421, 1335, 1192, 1086, 800, 721 cm^{-1}; ECD (MeOH) λ_{max} ($\Delta\varepsilon$) 341 (−2.37), 307 (+10.88), 272 (−10.88), 232 (+1.78) nm; ^{1}H and ^{13}C NMR data, see Table 1; HRESIMS *m/z* 265.1083 [M − H]$^-$ (calculated for $C_{14}H_{17}O_5$, 265.1081).

Raistrickione B (**2**): colorless plates (MeOH); mp 174–176 °C; $[\alpha]_D^{25}$ −48.6 (*c* 0.050, MeOH); UV (MeOH) λ_{max} (log ε) 220 (4.15), 280 (3.81) nm; IR (ATR) v_{max} 3224 (broad), 1678, 1596, 1423, 1336, 1094, 879, 798 cm^{-1}; ECD (MeOH) λ_{max} ($\Delta\varepsilon$) 342 (+2.62), 308 (−13.01), 273 (+14.20), 236 (−3.23) nm; ^{1}H and ^{13}C NMR data, see Table 1; HRESIMS *m/z* 265.1082 [M − H]$^-$ (calculated for $C_{14}H_{17}O_5$, 265.1081).

Raistrickione C (**3**): colorless powder; $[\alpha]_D^{25}$ +3.3 (*c* 0.058, MeOH); UV (MeOH) λ_{max} (log ε) 219 (4.21), 282 (3.89) nm; IR (ATR) v_{max} 3233 (broad), 1669, 1592, 1420, 1336, 1092, 998, 868, 800 cm^{-1}; ECD (MeOH) λ_{max} ($\Delta\varepsilon$) 342 (−2.86), 310 (+5.53), 276 (−7.55), 240 (+1.86) nm; ^{1}H and ^{13}C NMR data, see Table 1; HRESIMS *m/z* 267.1226 [M + H]$^+$ (calculated for $C_{14}H_{19}O_5$, 267.1227).

Raistrickione D (**4**): colorless needles (MeOH); mp 154–156 °C; $[\alpha]_D^{25}$ −66.8 (*c* 0.106, MeOH); UV (MeOH) λ_{max} (log ε) 220 (4.17), 286 (3.92) nm; IR (ATR) v_{max} 3337 (broad), 1668, 1588, 1417, 1333, 1080, 889, 788, 714 cm^{-1}; ECD (MeOH) λ_{max} ($\Delta\varepsilon$) 359 (+8.13), 321 (−17.14), 280 (+16.13) nm; ^{1}H and ^{13}C NMR data, see Table 2; HRESIMS *m/z* 279.1236 [M − H]$^-$ (calculated for $C_{15}H_{19}O_5$, 279.1238).

Raistrickione E (**5**): colorless needles; mp 180–182 °C; $[\alpha]_D^{25}$ −28.7 (*c* 0.080, MeOH); UV (MeOH) λ_{max} (log ε) 207 (4.32), 287 (3.97) nm; IR (ATR) v_{max} 3315 (broad), 3137 (broad), 1617, 1577, 1415, 1336, 1213, 1070, 891, 875, 755 cm^{-1}; ECD (MeOH) λ_{max} ($\Delta\varepsilon$) 327 (+13.62), 271 (−18.28), 227 (−0.78) nm; ^{1}H and ^{13}C NMR data, see Table 2; HRESIMS *m/z* 247.0976 [M − H]$^-$ (calculated for $C_{14}H_{15}O_4$, 247.0976).

4.4. X-ray Crystallographic Analysis of the Diastereomeric Mixture of **1** and **2**

$C_{14}H_{18}O_5$, *M* = 266.28, Monoclinic, space group *P*2(1); Unit cell dimensions were determined to be *a* = 11.7531(9) Å, *b* = 9.9251(6) Å, *c* = 11.7968(10) Å, α = 90°, β = 105.118 (2)°, γ = 90°, *V* = 1328.48

(17) Å^3, $Z = 4$, $D_{calculated} = 1.331$ mg/m^3, F(000) = 568, Crystal size 0.30 × 0.21 × 0.13 mm, μ(Cu Kα) = 0.840 mm^{-1}. Single crystals were measured on a Bruker Smart 1000 CCD X-ray diffractometer equipped with graphite-monochromated Cu Kα radiation ($\lambda = 1.54178$ Å) at 293 (2) K. A total of 7896 reflections were collected until $\theta_{max} = 66.04°$, in which 4137 independent unique reflections were observed ($R_{int} = 0.0712$). The structure was solved by direct methods with the SHELXTL software package, and refined by full-matrix least-squares on F^2. The final refinement gave $R_1 = 0.0628$ and $wR_2 = 0.1457$ (I > 2σ(I)). The crystallographic data for the structures of the diastereomeric mixture of **1** and **2** have been deposited in the Cambridge Crystallographic Data Centre (deposition number: CCDC 1839882).

4.5. Antioxidant Activity Assay

In the DPPH scavenging assay, the tested samples were dissolved in MeOH at the concentrations of 200 μM, 100 μM, 50 μM, 25 μM, and 12.5 μM. Then, 160 μL of the sample solutions was dispensed into the wells of a 96-well microtiter plate, and 40 μL of DPPH solutions in MeOH (400 μM, 200 μM, 100 μM, 50 μM, and 25 μM) was added to each well. The mixture was shaken vigorously and kept in the dark for 30 min. Then, the absorbance was measured at 517 nm using methanol as the blank reference. All of the experiments were performed in triplicate [31].

4.6. Cytotoxicity Assay

The cytotoxicity assay against human leukemia (HL60) cell lines was performed in triplicate using our previously described method, with doxorubicin as positive control (IC$_{50}$ value of 1.56 ± 0.32 μM). Five final concentrations (from 100 μM to 1 μM) in DMSO of the tested compound solutions were set in the wells of 96-well microtiter plates [31].

5. Conclusions

A thermo-change strategy applied to the prolific strain JH-18 of *P. raistrickii* afforded five new antioxidant PKs: raistrickiones A−E (**1–5**). At first, compounds **1** and **2** were obtained as a diastereomeric mixture, and their absolute configurations were determined in a crystal by X-ray diffraction analysis. Then, they were arduously separated by semi-preparative HPLC on a chiral column. Compounds **1–3** were diastereomeric at the C-8 and C-9 centers, but no distinct differences were observed in light of the antioxidant and cytotoxic results. In addition, compounds **1–5** showed considerable difference in skeleton with those reported previously from this fungal strain. Our work provides further demonstration that environmental cues such as huge thermo change represent a powerful approach in unlocking silent BCGs to produce new chemical compounds from fungi. The additional applications of other OSMAC strategies are currently underway, which could determine future reports on the discovery of new molecules from *P. raistrickii*.

Supplementary Materials: The following are available online at http://www.mdpi.com/1660-3397/16/6/213/s1, Figure S1. HRESIMS of raistrickione A (**1**), Figure S2: IR spectrum (ATR approach) of raistrickione A (**1**), Figure S3: UV spectrum (MeOH) of raistrickione A (**1**), Figure S4: ECD spectrum (MeOH) of raistrickione A (**1**), Figure S5: ^{1}H NMR spectrum (400 MHz DMSO-d_6) of raistrickione A (**1**), Figure S6: ^{13}C NMR spectrum (100 MHz DMSO-d_6) of raistrickione A (**1**), Figure S7: NOSEY spectrum (DMSO-d_6) of raistrickione A (**1**), Figure S8: HRESIMS of raistrickione B (**2**), Figure S9: IR spectrum (ATR approach) of raistrickione B (**2**), Figure S10:. UV spectrum (MeOH) of raistrickione B (**2**), Figure S11: ECD spectrum (MeOH) of raistrickione B (**2**), Figure S12: ^{1}H NMR spectrum (400 MHz DMSO-d_6) of raistrickione B (**2**), Figure S13:. ^{13}C NMR spectrum (100 MHz DMSO-d_6) of raistrickione B (**2**), Figure S14:. NOSEY spectrum (DMSO-d_6) of raistrickione B (**2**), Figure S15: ^{1}H NMR spectrum (400 MHz DMSO-d_6) of the diastereoisomeric mixture (**1** and **2**), Figure S16: ^{13}C NMR spectrum (100 MHz DMSO-d_6) of the diastereoisomeric mixture (**1** and **2**), Figure S17: DEPT of the diastereoisomeric mixture (**1** and **2**), Figure S18:. COSY of the diastereoisomeric mixture (**1** and **2**), Figure S19: HSQC of the diastereoisomeric mixture (**1** and **2**), Figure S20:. HMBC of the diastereoisomeric mixture (**1** and **2**), Figure S21: HRESIMS of raistrickione C (**3**), Figure S22:. IR of raistrickione C (**3**), Figure S23: UV spectrum (MeOH) of raistrickione C (**3**), Figure S24: ECD spectrum (MeOH) of raistrickione C (**3**), Figure S25: ^{1}H NMR spectrum (400 MHz DMSO-d_6) of raistrickione C (**3**), Figure S26: ^{13}C NMR spectrum (100 MHz DMSO-d_6) of raistrickione C (**3**), Figure S27: DEPT of raistrickione

C (**3**), Figure S28: COSY of raistrickione C (**3**), Figure S29: HSQC of raistrickione C (**3**), Figure S30: HMBC of raistrickione C (**3**), Figure S31: NOESY of raistrickione C (**3**), Figure S32: HRESIMS of raistrickione D (**4**), Figure S33: IR of raistrickione D (**4**), Figure S34: UV spectrum (MeOH) of raistrickione D (**4**), Figure S35: ^{1}H NMR spectrum (400 MHz acetone-d_6) of raistrickione D (**4**), Figure S36: ^{13}C NMR spectrum (100 MHz acetone-d_6) of raistrickione D (**4**), Figure S37: DEPT of raistrickione D (**4**), Figure S38: COSY of raistrickione D (**4**), Figure S39: HSQC of raistrickione D (**4**), Figure S40: HMBC of raistrickione D (**4**), Figure S41: NOESY of raistrickione D (**4**), Figure S42: HRESIMS of raistrickione E (**5**), Figure S43: IR of raistrickione E (**5**), Figure S44: UV spectrum (MeOH) of raistrickione E (**5**), Figure S45: ^{1}H NMR spectrum (400 MHz acetone-d_6) of raistrickione E (**5**), Figure S46: ^{13}C NMR spectrum (100 MHz acetone-d_6) of raistrickione E (**5**), Figure S47: DEPT of raistrickione E (**5**), Figure S48: COSY of raistrickione E (**5**), Figure S49: HSQC of raistrickione E (**5**), Figure S50: HMBC of raistrickione E (**5**), and the computational parts.

Author Contributions: W.-Z.L. designed the whole research work; D.-S.L., X.-G.R., H.-H.K., L.-Y.M. performed the experiments; D.-S.L. and W.-Z.L. analyzed the data and finished the paper; M.T.H. analyzed the data and helped to revise the paper.

Funding: This research was funded by National Natural Science Foundation of China (No. 31270082) and Natural Science Foundation of Shandong Province, China (No. Y2008B17 and ZR2013HM042).

Conflicts of Interest: The authors declare no conflict of interest.

References

1. Blunt, J.W.; Copp, B.R.; Keyzers, R.A.; Munro, M.H.G.; Prinsep, M.R. Marine natural products. *Nat. Prod. Rep.* **2017**, *34*, 235–294. [CrossRef] [PubMed]
2. Liu, D.S.; Yan, L.; Ma, L.Y.; Huang, Y.L.; Pan, X.H.; Liu, W.Z.; Lv, Z.H. Diphenyl derivatives from coastal saline soil fungus *Aspergillus iizukae*. *Arch. Pharm. Res.* **2015**, *38*, 1038–1043. [CrossRef] [PubMed]
3. Lin, Z.J.; Koch, M.; Abdel Aziz, M.H.; Galindo-Murillo, R.; Tianero, M.D.; Cheatham, T.E.; Barrows, L.R.; Reilly, C.A.; Schmidt, E.W. Oxazinin A, a pseudodimeric natural product of mixed biosynthetic origin from a filamentous fungus. *Org. Lett.* **2014**, *16*, 4774–4777. [CrossRef] [PubMed]
4. Williams, D.E.; Gunasekera, N.W.; Ratnaweera, P.B.; Zheng, Z.; Ellis, S.; Dada, S.; Patrick, B.O.; Wijesundera, R.L.C.; Nanayakkara, C.M.; Jefferies, W.A.; et al. Serpulanines A to C, N-oxidized tyrosine derivatives isolated from the Sri Lankan fungus *Serpula* sp.: Structure elucidation, synthesis, and histone deacetylase unhibition. *J. Nat. Prod.* **2018**, *81*, 78–84. [CrossRef] [PubMed]
5. Ren, J.W.; Niu, S.B.; Li, L.; Geng, Z.F.; Liu, X.Z.; Che, Y.S. Identification of oxaphenalenone ketals from the *Ascomycete* fungus *Neonectria* sp. *J. Nat. Prod.* **2015**, *78*, 1316–1321. [CrossRef] [PubMed]
6. Nielsen, J.C.; Grijseels, S.; Prigent, S.; Ji, B.; Dainat, J.; Nielsen, K.F.; Frisvad, J.C.; Workman, M.; Nielsen, J. Global analysis of biosynthetic gene clusters reveals vast potential of secondary metabolite production in *Penicillium* species. *Nat. Microbiol.* **2017**, *2*, 17044. [CrossRef] [PubMed]
7. Li, Y.F.; Tsai, K.J.S.; Harvey, C.J.B.; Li, J.J.; Ary, B.E.; Berlew, E.E.; Boehman, B.L.; Findley, D.M.; Friant, A.G.; Gardner, C.A.; et al. Comprehensive curation and analysis of fungal biosynthetic gene clusters of published natural products. *Fungal Genet. Biol.* **2016**, *89*, 18–28. [CrossRef] [PubMed]
8. Jiang, T.; Wang, M.; Li, L.; Si, J.; Song, B.; Zhou, C.; Yu, M.; Wang, X.; Zhang, Y.; Ding, G.; et al. Overexpression of the global regulator Laea in *Chaetomium globosum* leads to the biosynthesis of chaetoglobosin Z. *J. Nat. Prod.* **2016**, *79*, 2487–2494. [CrossRef] [PubMed]
9. Luo, Y.; Enghiad, B.; Zhao, H. New tools for reconstruction and heterologous expression of natural product biosynthetic gene clusters. *Nat. Prod. Rep.* **2016**, *33*, 174–182. [CrossRef] [PubMed]
10. Reen, F.J.; Romano, S.; Dobson, A.D.; O'Gara, F. The sound of silence: Activating silent biosynthetic gene clusters in marine microorganisms. *Mar. Drugs* **2015**, *13*, 4754–4783. [CrossRef] [PubMed]
11. Meng, L.H.; Liu, Y.; Li, X.M.; Xu, G.M.; Ji, N.Y.; Wang, B.G. Citrifelins A and B, citrinin adducts with a tetracyclic framework from cocultures of marine-derived isolates of *Penicillium citrinum* and *Beauveria felina*. *J. Nat. Prod.* **2015**, *78*, 2301–2305. [CrossRef] [PubMed]
12. He, X.Q.; Zhang, Z.Z.; Che, Q.; Zhu, T.J.; Gu, Q.Q.; Li, D.H. Varilactones and wortmannilactones produced by *Penicillium variabile* cultured with histone deacetylase inhibitor. *Arch. Pharm. Res.* **2018**, *41*, 57–63. [CrossRef] [PubMed]
13. Doull, J.L.; Singh, A.K.; Hoare, M.; Ayer, S.W. Conditions for the production of jadomycin B by *Streptomyces venezuelae* ISP5230: Effects of heat shock, ethanol treatment and phage infection. *J. Ind. Microbiol.* **1994**, *13*, 120–125. [CrossRef] [PubMed]

14. Wei, Z.H.; Wu, H.; Bai, L.Q.; Deng, Z.X.; Zhong, J.J. Temperature shift-induced reactive oxygen species enhanced validamycin A production in fermentation of *Streptomyces hygroscopicus* 5008. *Bioprocess Biosyst. Eng.* **2012**, *35*, 1309–1316. [CrossRef] [PubMed]

15. Ingebrigtsen, R.A.; Hansen, E.; Andersen, J.H.; Eilertsen, H.C. Light and temperature effects on bioactivity in diatoms. *J. Appl. Phycol.* **2016**, *28*, 939–950. [CrossRef] [PubMed]

16. Ma, L.Y.; Liu, D.S.; Li, D.G.; Huang, Y.L.; Kang, H.H.; Wang, C.H.; Liu, W.Z. Pyran rings containing polyketides from *Penicillium raistrickii*. *Mar. Drugs* **2017**, *15*, 2. [CrossRef] [PubMed]

17. Liu, W.Z.; Ma, L.Y.; Liu, D.S.; Huang, Y.L.; Wang, C.H.; Shi, S.S.; Pan, X.H.; Song, X.D.; Zhu, R.X. Peniciketals A–C, new spiroketals from saline soil derived *Penicillium raistrichii*. *Org. Lett.* **2014**, *16*, 90–93. [CrossRef] [PubMed]

18. Ma, L.Y.; Liu, W.Z.; Shen, L.; Huang, Y.L.; Rong, X.G.; Xu, Y.Y.; Gao, X.D. Spiroketals, isocoumarin and indoleformic acid derivatives from saline soil derived fungus *Penicillium raistrickii*. *Tetrahedron* **2012**, *68*, 2276–2282. [CrossRef]

19. Belofsky, G.N.; Gloer, K.B.; Gloer, J.B.; Wicklow, D.T.; Dowd, P.F. New *p*-terphenyl and polyketide metabolites from the sclerotia of *Penicillium raistrickii*. *J. Nat. Prod.* **1998**, *61*, 1115–1119. [CrossRef] [PubMed]

20. Pan, X.H.; Liu, D.S.; Wang, J.; Zhang, X.L.; Yan, M.M.; Zhang, D.H.; Zhang, J.J.; Liu, W.Z. Peneciraistin C induces caspase-independent autophagic cell death through mitochondrial-derived reactive oxygen species production in lung cancer cells. *Cancer Sci.* **2013**, *104*, 1476–1482. [CrossRef] [PubMed]

21. Bode, H.B.; Bethe, B.; Höfs, R.; Zeeck, A. Big effects from small changes: Possible ways to explore nature's chemical diversity. *Chembiochem* **2002**, *3*, 619–627. [CrossRef]

22. Karplus, M. Vicinal proton coupling in nuclear magnetic resonance. *J. Am. Chem. Soc.* **1963**, *85*, 2870–2871. [CrossRef]

23. Afifi, A.H.; Kagiyama, I.; El-Desoky, A.H.; Kato, H.; Mangindaan, R.E.P.; de Voogd, N.J.; Ammar, N.M.; Hifnawy, M.S.; Tsukamoto, S. Sulawesins A–C, furanosesterterpene tetronic acids that inhibit USP7, from a *Psammocinia* sp. marine sponge. *J. Nat. Prod.* **2017**, *80*, 2045–2050. [CrossRef] [PubMed]

24. Grkovic, T.; Pearce, A.N.; Munro, M.H.; Blunt, J.W.; Davies-Coleman, M.T.; Copp, B.R. Isolation and characterization of diastereomers of discorhabdins H and K and assignment of absolute configuration to discorhabdins D, N, Q, S, T, and U. *J. Nat. Prod.* **2010**, *73*, 1686–1693. [CrossRef] [PubMed]

25. Ragasa, C.Y.; de Luna, R.D.; Cruz, W.C., Jr.; Rideout, J.A. Monoterpene lactones from the seeds of *Nephelium lappaceum*. *J. Nat. Prod.* **2005**, *68*, 1394–1396. [CrossRef] [PubMed]

26. Evidente, A.; Andolfi, A.; Fiore, M.; Spanu, E.; Maddau, L.; Franceschini, A.; Marras, F.; Motta, A. Diplobifuranylones A and B, 5′-monosubstituted tetrahydro-2H-bifuranyl-5-ones produced by *Diplodia corticola*, a fungus pathogen of cork oak. *J. Nat. Prod.* **2006**, *69*, 671–674. [CrossRef] [PubMed]

27. Jarvis, B.B.; Wang, S. Stereochemistry of the roridins. diastereomers of roridin E. *J. Nat. Prod.* **1999**, *62*, 1284–1289. [CrossRef] [PubMed]

28. Presley, C.C.; Valenciano, A.L.; Fernández-Murga, M.L.; Du, Y.; Shanaiah, N.; Cassera, M.B.; Goetz, M.; Clement, J.A.; Kingston, D.G.I. Antiplasmodial chromanes and chromenes from the monotypic plant species *Koeberlinia spinosa*. *J. Nat. Prod.* **2018**, *81*, 475–483. [CrossRef] [PubMed]

29. Thoison, O.; Fahy, J.; Dumontet, V.; Chiaroni, A.; Riche, C.; Tri, M.V.; Sévenet, T. Cytotoxic prenylxanthones from *Garcinia bracteata*. *J. Nat. Prod.* **2000**, *63*, 441–446. [CrossRef] [PubMed]

30. Boonnak, N.; Chantrapromma, S.; Fun, H.K.; Yuenyongsawad, S.; Patrick, B.O.; Maneerat, W.; Williams, D.E.; Andersen, R.J. Three types of cytotoxic natural caged-scaffolds: Pure enantiomers or partial racemates. *J. Nat. Prod.* **2014**, *77*, 1562–1571. [CrossRef] [PubMed]

31. Ma, L.Y.; Liu, W.Z.; Huang, Y.L.; Rong, X.G. Two acid sorbicillin analogues from saline lands-derived fungus *Trichoderma* sp. *J. Antibiot.* **2011**, *64*, 645–647. [CrossRef] [PubMed]

marine drugs

MDPI

Review

Secondary Metabolites of Mangrove-Associated Strains of *Talaromyces*

Rosario Nicoletti [1,2], Maria Michela Salvatore [3] and Anna Andolfi [3,*

[1]   Council for Agricultural Research and Agricultural Economy Analysis, 00184 Rome, Italy; rosario.nicoletti@crea.gov.it
[2]   Department of Agriculture, University of Naples 'Federico II', 80055 Portici, Italy
[3]   Department of Chemical Sciences, University of Naples 'Federico II', 80125 Naples, Italy; mariamichela.salvatore@unina.it
*   Correspondence: andolfi@unina.it; Tel.: +39-081-2539179

Received: 8 November 2017; Accepted: 28 December 2017; Published: 6 January 2018

Abstract: Boosted by the general aim of exploiting the biotechnological potential of the microbial component of biodiversity, research on the secondary metabolite production of endophytic fungi has remarkably increased in the past few decades. Novel compounds and bioactivities have resulted from this work, which has stimulated a more thorough consideration of various natural ecosystems as conducive contexts for the discovery of new drugs. Thriving at the frontier between land and sea, mangrove forests represent one of the most valuable areas in this respect. The present paper offers a review of the research on the characterization and biological activities of secondary metabolites from manglicolous strains of species belonging to the genus *Talaromyces*. Aspects concerning the opportunity for a more reliable identification of this biological material in the light of recent taxonomic revisions are also discussed.

Keywords: bioactive products; drug discovery; endophytic fungi; mangroves; *Talaromyces*

---

1. Introduction

The establishment of the concept of 'one fungus, one name' in mycology [1] has stimulated reconsideration of the nomenclature of fungi, whose anamorphic stages were until recently grouped in the genus *Penicillium*, and included species renowned for being among the most prolific producers of bioactive secondary metabolites and a few blockbuster drugs [2–4]. In fact, a fundamental taxonomic revision has ultimately established that species with symmetrical biverticillate conidiophores, which were formerly ascribed to the *Penicillium* subgenus *Biverticillium*, are to be classified separately in the genus *Talaromyces*, and that *Penicillium* and *Talaromyces* belong to phylogenetic lineages that are distant enough to deserve ascription to different families [5,6]. Under the ecological viewpoint, recent reports are depicting a widespread endophytic occurrence of *Talaromyces* [7–10], which makes these fungi increasingly considered a source of interesting bioactive compounds.

After a few years, the above revision has not yet found full consideration. This is particularly true among researchers working in the field of drug discovery, who sometimes do not possess a robust mycological background. In fact, in a number of recent reports limiting identification to the genus level, the name *Penicillium* sp. is still inappropriately used for strains displaying the symmetrical biverticillate conidiophore condition. It is of course desirable that hasty investigators be more circumstantial in considering this fundamental step when reporting on their findings. Moreover, in contrast to the purpose of increasing accuracy, the adoption of identification procedures that are only based on DNA sequence homology has sometimes introduced additional approximation, considering that plenty of sequences referring to '*Penicillium* sp.' have been deposited in GenBank, and are routinely used as a support for the incomplete classification of new strains. Pending the diffusion of more decisive identification protocols, a good portion of the work carried out so far in the field of the purification and characterization of secondary metabolites from *Penicillium/Talaromyces* strains awaits revision in order to attain a more conclusive taxonomic ascription of this biological material, and avoid possible confusion from unreliable information. In fact, data concerning the production of secondary metabolites can be quite informative for these fungi, particularly when they are indicative of the ability to synthesize some structural models that are only, or predominantly, found in *Talaromyces* [11,12].

Following a recent paper on bioactive compounds from *Talaromyces* strains obtained from other marine sources [3], this review examines literature concerning secondary metabolites produced by these fungi recovered in association with mangrove plants, including a number of reports adopting the generic denomination '*Penicillium* sp.'.

2. Mangrove Swamps: A Dynamic Frontier between Land and Sea

Spread along the coastlines at tropical and subtropical latitudes, mangrove forests are a biodiversity hotspot as well as a peculiar transition ecosystem, harboring organisms that are typical of either marine or terrestrial habitats. Considering their prevalently emerged bearing, mangrove plants cannot be considered real marine organisms to the same extent as seagrasses [3]. However, they play a key role in maintaining and building soil from the intertidal zone, and are morphologically and physiologically adapted to the particularly harsh environmental conditions deriving from a combination of extensive salinity, tide alternation, anaerobic clayey soil, high temperature, and moisture [13].

Mangrove plants host a great variety of endophytic and other associated fungi, a good part of which derives from the surrounding soil, marine, and freshwater contexts. Regardless of their true origin, which in most instances cannot be proven, these symbionts might contribute to their host's adaptation in such a peculiar habitat [14]. According to the plant species, the environmental conditions, and other factors, a wide set of interactions are potentially established between endophytes and their hosts [15]. However, the most considered aspect is represented by the mutual effects on the production of secondary metabolites. Recent investigations have demonstrated that these secondary metabolites are regulated by complex biomolecular mechanisms, such as chromatin methylation [16], and are regarded as fundamental mediators of interspecific communication [17]. In applicative terms, this intriguing ecological scenario reflects a series of bioactive properties of a multitude of structurally diverse compounds that these fungi are able to synthesize, stimulating their consideration as one of the most promising sources for drug prospects [14,18–22].

3. The Occurrence of *Talaromyces* Species in Mangroves

In the last decade, literature concerning drug discovery has been substantially enriched by many reports dealing with the biosynthetic potential of mangrove-associated fungi. Also, there has been an increasing trend over the past few years in the finding of *Talaromyces* strains from this particular ecological context, which appears to be in evident connection with its quite recent spread in nomenclatural use following the formal separation from *Penicillium*. However, apart from two cases from South America, these reports all refer to locations in southeast Asia, particularly from the Chinese provinces of Fujian, Guangdong, and Guangxi, and Hainan Island (Table 1).

Table 1. List of mangrove-associated *Talaromyces* strains gathered from the literature.

Species/Strain	Source	Location	Reference
T. aculeatus/9EB	*Kandelia candel* (leaf)	Yangjiang (Guangdong), China	[23]
T. amestolkiae/YX1	*Kandelia obovata* (leaf)	Zhanjiang Mangrove Natural Reserve (Guangdong), China	[24]
T. amestolkiae/HZ-YX1	*K. obovata* (leaf)	Huizhou Mangrove Natural Reserve (Guangdong), China	[25]

Table 1. *Cont.*

Species/Strain	Source	Location	Reference
T. atroroseus/IBT 20955	*Laguncularia racemosa* (root)	Paria Bay, Venezuela	[26]
T. flavus/CCTCCM2010266	*Sonneratia apetala* (leaf)	Hainan, China	[27]
T. funiculosus	*Avicennia officinalis* (root) *Rhizophora mucronata* (root) undetermined species (leaf)	Pichavaram (Tamil Nadu), India	[28]
T. pinophilus/HN29-3B1	*Cerbera manghas*	Dong Zhai Gang Mangrove Natural Reserve (Hainan), China	[29]
T. pinophilus	*Ceriops tagal* (root)	Dong Zhai Gang (Hainan), China	[30]
T. pinophilus	*L. racemosa* (leaf)	Itamaracá Island, Brazil	[31]
T. purpurogenus/JP-1	*Aegiceras corniculatum* (bark)	Fujian, China	[32]
Talaromyces sp./FJ-1 [1]	*C. tagal* (stem)	Haikou (Hainan), China	[33]
Talaromyces sp./FJ-1 [1]	*Avicennia marina*	Fujian, China	[34]
Talaromyces sp./FJ-1 [1]	*Acanthus ilicifolius*	Hainan, China	[35]
Talaromyces sp./ZJ-SY2 [1]	*S. apetala* (leaf)	Zhanjiang Mangrove Natural Reserve (Guangdong), China	[36]
Talaromyces sp./SBE-14	*K. candel* (bark)	Hong Kong, China	[37]
Talaromyces sp./ZH154	*K. candel* (bark)	Zhuhai (Guangdong), China	[38]
T. stipitatus/SK-4	*A. ilicifolius* (leaf)	Shankou Mangrove Natural Reserve (Guangxi), China	[39]
T. trachyspermus/KUFA35	not specified	Thailand	[40]

[1] These strains reported as *Penicillium* sp.

However, it is questionable whether some of these reports are actually replications. In fact, the strains YX1 and HZ-YX1 obtained from leaf samples of *Kandelia obovata* were claimed to have been collected in April 2012 at two locations in the Guangdong province situated over 400 km apart. Both strains were ascribed to the species *T. amestolkiae* based on rDNA-ITS sequence homology; nevertheless, the same GenBank accession code is indicated by the authors, which refers to Zhanjiang as the place of origin (hence strain YX1) [24,25]. Even more ambiguous is the case of strain 9EB of *T. aculeatus*, whose identification was again based on the homology of a 16S sequence of 576 bp deposited in GenBank (accession code: KT715695), which is actually referred to a strain of *Penicillium* sp. that had been given a different number (C08652) [23]. However, this sequence is identical to one from another strain (CY196, accession number: KP059103) identified as *T. verruculosus*, again submitted from Chinese researchers from Guangzhou. Finally, substantial perplexity arises for three strains labeled with the same number (FJ-1) despite a declared different origin, which are reported to have been identified through rDNA-ITS sequencing [33–35]. However, the GenBank code (DQ365947.1) provided for all of them actually corresponds to a previously deposited sequence from a strain of *T. purpurogenus* (HS-A82).

4. Structures and Properties of Secondary Metabolites from Manglicolous *Talaromyces*

Most of the strains mentioned in Table 1 were reported for the production/bioactive effects of secondary metabolites, which undoubtedly represent the major objective prompting research on endophytic fungi. The structure of these compounds was essentially elucidated by means of spectroscopic methods, such as two-dimensional (2D) NMR and mass spectrometry. In some cases, their absolute configuration was determined through a modified Mosher's method or electronic circular dichroism (ECD) spectra, or the structures confirmed by means of single-crystal X-ray diffraction experiments. So far, 39 new compounds out of a total of 88 (Table 2) have resulted from the biochemical characterization of these strains. Aside from a few quite original structural models, most of them are strictly correlated to known products that have been previously reported from other strains of *Talaromyces* [2,3,11]. A lower number of compounds (22) already known from this genus have also been identified in manglicolous strains, indicating that research in this particular field has yielded a notable percentage of new products. However, it is not possible to infer whether these numbers subtend any specific biosynthetic abilities, considering that it is quite likely that a few novel products were not previously detected in strains of different origin by the simple reason that they had not been characterized yet.

Table 2. Structures and bioactivities of secondary metabolites produced by manglicolous *Talaromyces* strains. The names of novel compounds are underlined. Compounds marked by an asterisk were previously reported from *Talaromyces* strains from sources other than mangroves [3,11,12].

Compound Name	Structure	Reported Bioactivities	Reference
Depsidones, Diphenyl Ether Derivatives			
Penicillide * (R = H) Purpactin A * (=vermixocin B) (R = CH₃CO)			[39]
Secopenicillide B			[39]
Talaromyone A (R = H)		Antibacterial: (MIC µg/mL) *B. subtilis* 12.5 (talaromyone B)	[39]
Talaromyone B (R = CH₃CO)		α-Glucosidase inhibitor (IC₅₀ µM) 48.4 (talaromyone B)	
Tenelate A (R = H) Tenelate B (R = CH₂CH₃)			[37]
Tenellic acid A *		α-Glucosidase inhibitor (IC₅₀ µM) 99.8	[39]
Tenellic acid C			[37,39]
Funicones, Vermistatins			
3-O-Methylfunicone *			[41]
Penicidone D			[41]
(±)-Penifupyrone		α-Glucosidase inhibitor (IC₅₀ µM) 14.4	[41]

Table 2. *Cont.*

Compound Name	Structure	Reported Bioactivities	Reference
Penisimplicissin * (R$_1$ = H, R$_2$ = CH$_3$) 6-Demethylpenisimplicissin (R$_1$ = R$_2$ = H) 5′-Hydroxypenisimplicissin (R$_1$ = OH, R$_2$ = CH$_3$)		α-Glucosidase inhibitor (IC$_{50}$ μM) 9.5 (6-demethylpenisimplicissin)	[29]
Vermistatin * (R = H) Hydroxyvermistatin * (R = OH) Methoxyvermistatin * (R = OCH$_3$)		α-Glucosidase inhibitors (IC$_{50}$ μM) 29.2, 20.3 [1]	[29]
2″-Epihydroxydihydrovermistatin		α-Glucosidase inhibitor (IC$_{50}$ μM) 8.0	[29]
6-Demethylvermistatin			[29]

Anthraquinones			
Emodin *		Antibacterial: (MIC μg/mL) *E. coli* 6.25; *P. aeruginosa* 12.5; *S. ventriculi* 12.5; *S. aureus* 12.5 Antifungal: (MIC μg/mL) *A. niger* 12.5; *C. albicans* 6.25; *F. oxysporum* f.sp. *cubense* 25.0 Cytotoxic: (IC$_{50}$ μg/mL) KB 12.43; KBv200 15.72	[38]
Skyrin *		Antibacterial: (MIC μg/mL) *E. coli* 25.0; *P. aeruginosa* 12.5; *S. ventriculi* 25.0; *S. aureus* 25.0 Antifungal: (MIC μg/mL) *A. niger* 25.0; *C. albicans* 12.25 Cytotoxic: (IC$_{50}$ μg/mL) KB 20.38; KBv200 16.06	[38]

Xanthones			
Conioxanthone A (R$_1$ = H, R$_2$ = R$_3$ = OH) 8-Hydroxy-6-methyl-9-oxo-9H-xanthene-1-methylcarboxylate (R$_1$ = R$_2$ = R$_3$ = H) Pinselin (R$_1$ = OH, R$_2$ = R$_3$ = H) Sydowinin A (R$_1$ = OH, R$_2$ = H, R$_3$ = OH) Sydowinin B (R$_1$ = R$_2$ = H, R$_3$ = OH)		Immunosuppressive: (IC$_{50}$ μg/mL) Con A-Induced 8.2, 25.7, 5.9, 6.5, 19.2 [1] Immunosuppressive: (IC$_{50}$ μg/mL) LPS-Induced 7.5, 26.4, 7.5, 7.1, 20.8 [1]	[36]
Norlichexanthone		Antibacterial: (MIC μg/mL) *P. aeruginosa* 25.0; *S. ventriculi* 25.0; *S. aureus* 12.5 Antifungal (MIC μg/mL) *A. niger* 25.0; *C. albicans* 6.25; *F. oxysporum* f.sp. *cubense* 50.0 Cytotoxic: (IC$_{50}$ μg/mL) KB 12.43; KBv200 15.72	[38]
Peniphenone (R = H)		Immunosuppressive: Con A-Induced (IC$_{50}$ μg/mL) 8.1, 17.5 [1]	[36]
Methylpeniphenone (R = CH$_3$)		Immunosuppressive: LPS-Induced (IC$_{50}$ μg/mL) 9.3, 23.7 [1]	

Table 2. *Cont.*

Compound Name	Structure	Reported Bioactivities	Reference
Remisporine B (R = βH) Epiremisporine B (R = αH)			[36]
Secalonic acid A		Antibacterial: (MIC μg/mL) *E. coli* 25.0; *P. aeruginosa* 12.5; *S. ventriculi* 12.5; *S. aureus* 12.5 Antifungal (MIC μg/mL) *A. niger* 6.25; *C. albicans* 6.25; *F. oxysporum* f.sp. *cubense* 12.5 Cytotoxic: (IC$_{50}$ μg/mL) KB 0.63; KBv200 1.05	[38]
Stemphyperylenol		Antibacterial: (MIC μg/mL) *P. aeruginosa* 12.5; *S. ventriculi* 3.12; *S. aureus* 25.0 Antifungal: (MIC μg/mL) *A. niger* 50.0; *C. albicans* 6.25 Cytotoxic: (IC$_{50}$ μg/mL) KB 20.20; KBv200 44.35	[38]
Benzophenone Analogs			
Arugosin I			[32]
Penicillenone		Cytotoxic: (IC$_{50}$ μM) P388 1.38	[32]
Phenols, Biphenyls			
4-(2′,3′-Dihydroxy-3′-ethyl-butanoxy)-phenethanol		Cytotoxic: (IC$_{50}$ μM) MG-63 35, Tca8113 26	[34]
2,4-Dihydroxy-6-methylbenzoic acid (R = COOH) 5-Methylbenzene-1,3-diol (R = H)			[41]
4′-(*S*)-(3,5-Dihydroxyphenyl)-4′-hydroxy-6′-methylcyclopent-1′-en-5′-one			[41]
6′-Methyl-[1,1′-biphenyl]-3,3′,4′,5-tetraol		α-Glucosidase inhibitor (IC$_{50}$ μM) 2.2	[41]
Benzofurans			
5-Carboxyphthalide			[23]

Table 2. *Cont.*

Compound Name	Structure	Reported Bioactivities	Reference
1-(5-Hydroxy-7-methoxy-benzofuran-3-yl)-ethanone		Antibacterial: (MIC µg/mL) *B. subtilis* 50; *E. coli* 50; *S. aureus* 25; *S. epidermidis* 50	[24]
5-Hydroxy-7-methoxy-2-methyl-benzofuran-3-carboxylic acid		Antibacterial (MIC µg/mL) *B. subtilis* 25; *E. coli* 50; *S. aureus* 25; *S. epidermidis* 25	[24]
Isocoumarins			
Aspergillumarin A *		α-Glucosidase inhibitor (IC$_{50}$ µM) 38.1	[24]
Aspergillumarin B * ($R_1 = R_2 = H$) Penicimarin B * ($R_1 = CH_3$, $R_2 = H$) Penicimarin C * ($R_1 = CH_3$, $R_2 = OH$)		α-Glucosidase inhibitors (IC$_{50}$ µM) 193.1, 431.4, 266.3 [1]	[24]
6,8-Dihydroxy-3,4-dimethyl-isocoumarin ($R_1 = H$, $R_2 = H$, $R_3 = CH_3$) 6,8-Dihydroxy-5-methoxy-3-methyl-isochromen-1-one ($R_1 = H$, $R_2 = CH_3$, $R_3 = H$) 6-Hydroxy-8-methoxy-3,4-dimethylisocoumarin ($R_1 = CH_3$, $R_2 = H$, $R_3 = CH_3$)		α-Glucosidase inhibitor (IC$_{50}$ µM) 34.4, 89.4, 585.7 [1]	[24]
3-(4,5-Dihydroxy-pentyl)-8-hydroxy-isochroman-1-one		α-Glucosidase inhibitor (IC$_{50}$ µM) 162.5	[24]
5,6-Dihydroxy-3-(4-hydroxy-pentyl)-isochroman-1-one		α-Glucosidase inhibitor (IC$_{50}$ µM) 142.1	[24]
6-Hydroxy-4-(1-hydroxy-ethyl)-8-methoxy-isocoumarin [2] ($R_1 = CH_3$, $R_2 = H$) Sescandelin ($R_1 = R_2 = H$) 5,6,8-Trihydroxy-4-(1-hydroxy-ethyl)-isocoumarin ($R_1 = H$, $R_2 = OH$)		α-Glucosidase inhibitor (IC$_{50}$ µM) 537.3	[24]
6-Hydroxy-4-hydroxymethyl-8-methoxy-3-methyl-isocoumarin ($R = CH_3$) Sescandelin B * ($R = H$)		α-Glucosidase inhibitors (IC$_{50}$ µM) 302.6, 17.2 [1]	[24]
Isobutyric acid 5,7-dihydroxy-2-methyl-4-oxo-3,4-dihydro-naphththalen-1-yl methyl ester		α-Glucosidase inhibitor (IC$_{50}$ µM) 140.8	[24]
Deoxytalaroflavone		Antibacterial (*S. aureus*)	[33]

Table 2. Cont.

Compound Name	Structure	Reported Bioactivities	Reference
7-Hydroxy-deoxytalaroflavone		Antibacterial (*S. aureus*, m.r.-*S. aureus*)	[33]
Azaphilones			
7-Epiaustdiol (R = H) 8-*O*-Methylepiaustdiol (R = CH₃)		Antibacterial: (MIC μg/mL) *E. coli* > 100, 25; *P. aeruginosa* 6.26, 25.0; *S. ventriculi* 25.0, 50; *S. aureus* 12.6, 50.0 [1] Antifungal: (MIC μg/mL) *A. niger* 25.0, 50.0; *C. albicans* 12.5, 25.0 [1] Cytotoxic: (IC₅₀ μg/mL) KB 20.04, 16.37; KBv200 19.32, 37.16 [1]	[38]
Monascorubramine			[26]
Monascorubrin			[26]
Pinazaphilone A		α-Glucosidase inhibitor (IC₅₀ μM) 81.7	[41]
Pinazaphilone B (R₁ = CH₃, R₂ = OH) Sch 1385568 * (R₁ = OH, R₂ = CH₃)		α-Glucosidase inhibitor (IC₅₀ μM) 28.0	[41]
Sequoiamonascin D			[32]
Sequoiatone A			[32]
Sequoiatone B			[32]

Table 2. *Cont.*

Compound Name	Structure	Reported Bioactivities	Reference
Nonadrides			
Glauconic acid *			[26]
Phenalenone Derivatives			
Bacillosporin A * (R = CH₃CO) Bacillosporin B * (R = H)		α-Glucosidase inhibitors (IC$_{50}$ μM) 33.55, 95.81 [1]	[23,32]
Bacillosporin C *			[32]
9-Demethyl FR-901235			[32]
Chromones			
(2′S *)-2-(2′-Hydroxypropyl)-5-methyl-7,8-dihydroxy-chromone		Antibacterial (MIC μM) *Salmonella* 2.0	[23]
Cyclohexenones			
Leptosphaerone C		Cytotoxic: (IC$_{50}$ μM) A-549 1.45	[32]
Flavonoids			
(2R,3S)-Pinobanksin-3-cinnamate		Neuroprotective	[35]
Alkaloids			
Talaramide		Antimycobacterial: (IC$_{50}$ μM) PknG kinase inhibitor 55	[25]
ZG-1494α *			[26]

Table 2. *Cont.*

Compound Name	Structure	Reported Bioactivities	Reference
Terpenes			
15-Hydroxy-6α,12-epoxy-7β,10αH,11βH-spiroax-4-ene-12-one		Cytotoxic: (IC$_{50}$ μM) MG-63 55nM, Tca8113 10, WRL-68 58	[34]
15-α-Hydroxy-(22E,24R)-ergosta-3,5,8(14),22-tetraen-7-one		Cytotoxic: glioma cell lines (IC$_{50}$ μM) U251 3.2, BT-325 4.1, SHG-44 2.3	[35]
Purpuride *			[26]
Steperoxide B (=merulin A) (R = H)		Toxic to brine shrimp	[27]
Talaperoxide A (R = CH$_3$CO)		Cytotoxic: (IC$_{50}$ μM) HeLa 7.97, 13.7; HepG2 6.79, 12.93; MCF-7 4.17, 19.77; MDA-MB-435 1.90, 11.78; PC-3 1.82, 5.70 [1]	
Talaperoxide B		Toxic to brine shrimp Cytotoxic: (IC$_{50}$ μM) HeLa 1.73, HepG2 1.29; MCF-7 1.33; MDA-MB-435 2.78; PC-3 0.89	[27]
Talaperoxide C		Toxic to brine shrimp Cytotoxic: (IC$_{50}$ μM) HeLa 12.71; HepG2 15.11; MCF-7 6.63; MDA-MB-435 2.64; PC-3 4.34	[27]
Talaperoxide D		Toxic to brine shrimpCytotoxic: (IC$_{50}$ μM) HeLa 1.31; HepG2 0.90; MCF-7 1.92; MDA-MB-435 0.91; PC-3 0.70	[27]

[1] Data were reported according to the order of compounds; [2] This compound was incorrectly named 5-hydroxy-4-(1-hydroxy-ethyl)-8-methoxy-isocoumarin in the original report.

The majority of these secondary metabolites have been evaluated for some kind of biological properties, particularly cytotoxic/antiproliferative activity against tumor cell lines, antimicrobial effects against bacterial and fungal strains, and immunosuppressive and enzyme inhibitory aptitudes. However, some interesting effects have been also described for many of the other 49 compounds previously reported from other biological sources, which have not been specifically considered in Table 2.

As a likely result of evolutionary pressure, genes encoding fungal secondary metabolites are known to be clustered, and their synthesis is known to occur through a few common schemes, such as the acetate, shikimate, and mevalonate pathways [42]. Nevertheless, the molecular structure of these compounds is very varied, even within a single genus such as *Talaromyces*, and a convenient discussion should be based on their grouping in different classes [43].

Depsidones are ester-like depsides, or cyclic ethers, which are related to the diphenyl ethers, and synthesized through the polymalonate pathway. Their structure is based on an 11*H*-dibenzo(*b*,*e*) [1,4] dioxepin-11-one ring system where bridging at the phenolic group in the *p*-position can result in increased antioxidant activity. The efficient antioxidant properties of depsidones may also derive from their incorporation into lipid microdomains [44]. Since antioxidant properties are in turn related to anti-inflammatory, antiproliferative, and antiviral activities, compounds from *Talaromyces* spp. belonging to this class, particularly the novel talaromyones A and B [39], should be better investigated with reference to these bioactive effects. Funicones and the related vermistatins probably represent the most typical class of secondary metabolites produced by *Talaromyces* spp., possessing several bioactive properties that make them renowned drug prospects [45]. Particularly, 3-*O*-methylfunicone has displayed notable antifungal, antitumor, and lipid-lowering properties that require more circumstantial investigations beyond academic research, for which a direct support by the pharmaceutical industry seems to be fundamental [46–51]. A few novel vermistatin derivatives obtained from a manglicolous strain of *T. pinophilus* have been characterized as α-glucosidase inhibitors [29].

In fungi, both anthraquinones and xanthones are reported to be synthesized through the cyclization of polyacetate units, in the latter case followed by oxidative cleavage of the central ring [52]. Well-known mycotoxins ascribed to these groups, such as emodin, skyrin, secalonic acid A, and norlichexanthone, have been also reported as secondary metabolites of a manglicolous *Talaromyces* strain [38]. The related benzophenones are represented by the new potent immunosuppressive product peniphenone and its methyl derivative [32]. Other phenolic metabolites are possibly synthesized following the shikimate pathway [53], such as two new biphenyl and phenylcyclopentenone derivatives that have been characterized for their α-glucosidase inhibitory effects [41].

Benzofurans, also known as coumarones, represent another important class of natural products, and a scaffold considered for the development of synthetic drugs [54]. This group includes two new compounds derived from strain YX1 of *T. amestolkiae*, which exhibit antibacterial activities [24]. Again known for a wide array of pharmacological properties, isocoumarins are coumarin isomers presenting an inverted lactone ring, most of which possess a 3-alkyl or a 3-phenyl moiety on a α-pyranone nucleus, and 8-oxygenation on the benzene ring. The discovery of novel natural isocoumarins is ongoing; a few hundreds of isocoumarins and dihydroisocoumarins are currently known from different sources [55]. Despite such a high diversity, the number of isocoumarins displaying a completely different substitution pattern is quite reduced, and most of the newly isolated products turn out to be derivatives of previously known structures. A good example is represented by a series of known and novel compounds by the above-mentioned strain YX1, which have been again characterized for their α-glucosidase inhibitory effects [24]. The talaroflavones, including the new antibacterial analogue 7-hydroxy-deoxytalaroflavone [33], are also ascribed to this class.

Azaphilones are a typical class of fungal red or purple pigments with pyrone–quinone structures containing a highly oxygenated bicyclic core and a chiral quaternary center, whose use as colorants has been proposed in several fields, including the food industry [56]. These compounds exhibit a wide range of bioactivities, deriving from antimicrobial, antiviral, antioxidant, cytotoxic, nematicidal, and anti-inflammatory properties [57,58]. New members of this family are represented by the antibacterial/cytotoxic product 7-epiaustdiol and its methyl derivative [38], and the pinazaphilones, which have been characterized as α-glucosidase inhibitors [41]. Another red pigment, glauconic acid, is probably the oldest product mentioned in this review. In fact, this nonadride compound has been known since 1931 [59], mainly from studies concerning its biosynthetic pathway, which indicate that it derives through several steps involving substitutions in citric acid and dimerization of a C_9 anhydride unit [60], or from succinate [61]. However, no detailed investigation of its bioactivity seems to have been accomplished so far. α-glucosidase inhibitory activity also characterizes bacillosporins (bacillisporins) A and B, two known antibacterial oligophenalenone dimers reported together with a new chromone from a strain of *T. aculeatus* [23]. Additional novel polyketides from manglicolous *Talaromyces* strains are represented by leptosphaerone C, a cytotoxic cyclohexenone derivative [32], and the flavonoid (2*R*,3*S*)-pinobanksin-3-cinnamate, displaying interesting neuroprotective effects [35].

Although alkaloids are widespread secondary metabolites of endophytic fungi [62], only two representatives of this class have been reported from mangrove-associated *Talaromyces* strains. Particularly, ZG-1494α is a pyrrolidinone derivative that has been reported as an inhibitor of the platelet-activating factor acetyltransferase [63], while talaramide A is a new compound presenting an unusual oxidized tricyclic system, which has been characterized for its antimycobacterial properties deriving from PknG kinase inhibitory effects [25].

Finally, the terpenes also appear to be quite infrequent from this particular microbial source. They include the sesquiterpene amino acid-alcohol ester purpuride [26], and a few novel cytotoxic-antiproliferative products, namely 15-hydroxy-6α,12-epoxy-7β,10αH,11βH-spiroax-4-ene-12-one [34], 15-α- hydroxy-(22E,24R)-ergosta-3,5,8(14),22-tetraen-7-one [35], and the talaperoxide series [27].

5. Conclusions

The availability of increasingly refined laboratory equipment, and the ability to access previously hindered sources for the isolation of novel fungal strains has stimulated a huge amount of research activity in view of identifying new bioactive compounds and drugs. Moreover, novel accurate screening strategies and procedures have been introduced for a targeted selection in view of reducing the misuse of resources and ensuing replication through finding known compounds [64–67]. With an increasing rate of recovery from both terrestrial and marine environmental contexts, and a wide range of ecological interactions with other organisms, *Talaromyces* strains are among the most promising 'biofactories' that can further enlarge the current panorama of bioactive products available for exploitation by the pharmaceutical industry. Considering the relatively reduced extension of the areas covered by such investigations so far, this remarkable potential deserves to be more thoroughly appreciated, particularly by spreading the search for new strains all over the manglicolous regions that have not yet been considered.

Author Contributions: R.N. and A.A. conceived and organized the manuscript, and wrote the text; M.M.S. gathered data from the literature and prepared Table 2.

Conflicts of Interest: The authors declare no conflict of interest.

References

1. Hawksworth, D.L.; Crous, P.W.; Redhead, S.A.; Reynolds, D.R.; Samson, R.A.; Seifert, K.A.; Taylor, J.W.; Wingfield, M.J.; Abaci, O.; Aime, C.; et al. The Amsterdam declaration on fungal nomenclature. *IMA Fungus* **2011**, *2*, 105–112. [CrossRef] [PubMed]

2. Frisvad, J.C. Taxonomy, chemodiversity, and chemoconsistency of *Aspergillus*, *Penicillium*, and *Talaromyces* species. *Front. Microbiol.* **2015**, *5*, 773. [CrossRef] [PubMed]

3. Nicoletti, R.; Trincone, A. Bioactive compounds produced by strains of *Penicillium* and *Talaromyces* of marine origin. *Mar. Drugs* **2016**, *14*, 37. [CrossRef] [PubMed]

4. Koul, M.; Singh, S. *Penicillium* spp.: Prolific producer for harnessing cytotoxic secondary metabolites. *Anti-Cancer Drugs* **2017**, *28*, 11–30. [CrossRef] [PubMed]

5. Houbraken, J.; Samson, R.A. Phylogeny of *Penicillium* and the segregation of *Trichocomaceae* into three families. *Stud. Mycol.* **2011**, *70*, 1–51. [CrossRef] [PubMed]

6. Yilmaz, N.; Visagie, C.M.; Houbraken, J.; Frisvad, J.C.; Samson, R.A. Polyphasic taxonomy of the genus *Talaromyces*. *Stud. Mycol.* **2014**, *78*, 175–341. [CrossRef] [PubMed]

7. Li, L.Q.; Yang, Y.G.; Zeng, Y.; Zou, C.; Zhao, P.J. A new azaphilone, kasanosin C, from an endophytic *Talaromyces* sp. T1BF. *Molecules* **2010**, *15*, 3993–3997. [CrossRef] [PubMed]

8. Bara, R.; Aly, A.H.; Pretsch, A.; Wray, V.; Wang, B.; Proksch, P.; Debbab, A. Antibiotically active metabolites from *Talaromyces wortmannii*, an endophyte of *Aloe vera*. *J. Antibiot.* **2013**, *66*, 491–493. [CrossRef] [PubMed]

9. Palem, P.P.; Kuriakose, G.C.; Jayabaskaran, C. An endophytic fungus, *Talaromyces radicus*, isolated from *Catharanthus roseus*, produces vincristine and vinblastine, which induce apoptotic cell death. *PLoS ONE* **2015**, *10*, e0144476. [CrossRef] [PubMed]

10. Vinale, F.; Nicoletti, R.; Lacatena, F.; Marra, R.; Sacco, A.; Lombardi, N.; d'Errico, G.; Digilio, M.C.; Lorito, M.; Woo, S.L. Secondary metabolites from the endophytic fungus *Talaromyces pinophilus*. *Nat. Prod. Res.* **2017**, *31*, 1778–1785. [CrossRef] [PubMed]

11. Frisvad, J.C.; Filtenborg, O.; Samson, R.A.; Stolk, A.C. Chemotaxonomy of the genus *Talaromyces*. *Antonie van Leeuwenhoek* **1990**, *57*, 179–189. [CrossRef] [PubMed]

12. Zhai, M.M.; Li, J.; Jiang, C.X.; Shi, Y.P.; Di, D.L.; Crews, P.; Wu, Q.X. The bioactive secondary metabolites from *Talaromyces* species. *Nat. Prod. Bioprospect.* **2016**, *6*, 1–24. [CrossRef] [PubMed]

13. Lee, S.Y.; Primavera, J.H.; Dahdouh-Guebas, F.; McKee, K.; Bosire, J.O.; Cannicci, S.; Diele, K.; Fromard, F.; Koedam, N.; Marchand, C.; et al. Ecological role and services of tropical mangrove ecosystems: A reassessment. *Glob. Ecol. Biogeogr.* **2014**, *23*, 726–743. [CrossRef]

14. Debbab, A.; Aly, A.H.; Proksch, P. Mangrove derived fungal endophytes—A chemical and biological perception. *Fungal Divers.* **2013**, *61*, 1–27. [CrossRef]

15. Wani, Z.A.; Ashraf, N.; Mohiuddin, T. Plant-endophyte symbiosis, an ecological perspective. *Appl. Microbiol. Biotechnol.* **2015**, *99*, 2955–2965. [CrossRef] [PubMed]

16. Chujo, T.; Scott, B. Histone H3K9 and H3K27 methylation regulates fungal alkaloid biosynthesis in a fungal endophyte–plant symbiosis. *Mol. Microbiol.* **2014**, *92*, 413–434. [CrossRef] [PubMed]

17. Netzker, T.; Fischer, J.; Weber, J.; Mattern, D.J.; König, C.C.; Valiante, V.; Schroeckh, V.; Brakhage, A.A. Microbial communication leading to the activation of silent fungal secondary metabolite gene clusters. *Front. Microbiol.* **2015**, *6*, 299. [CrossRef] [PubMed]

18. Cheng, Z.S.; Pan, J.H.; Tang, W.C.; Chen, Q.J.; Lin, Y.C. Biodiversity and biotechnological potential of mangrove-associated fungi. *J. For. Res.* **2009**, *20*, 63–72. [CrossRef]

19. Debbab, A.; Aly, A.H.; Proksch, P. Bioactive secondary metabolites from endophytes and associated marine derived fungi. *Fungal Divers.* **2011**, *49*, 1. [CrossRef]

20. Thatoi, H.; Behera, B.C.; Mishra, R.R. Ecological role and biotechnological potential of mangrove fungi: A review. *Mycology* **2013**, *4*, 54–71.

21. Wang, X.; Mao, Z.G.; Song, B.B.; Chen, C.H.; Xiao, W.W.; Hu, B.; Wang, J.W.; Jiang, X.B.; Zhu, Y.H.; Wang, H.J. Advances in the study of the structures and bioactivities of metabolites isolated from mangrove-derived fungi in the South China Sea. *Mar. Drugs* **2013**, *11*, 3601–3616. [CrossRef] [PubMed]

22. Wang, K.W.; Wang, S.W.; Wu, B.; Wei, J.G. Bioactive natural compounds from the mangrove endophytic fungi. *Mini Rev. Med. Chem.* **2014**, *14*, 370–391. [CrossRef] [PubMed]

23. Huang, H.; Liu, T.; Wu, X.; Guo, J.; Lan, X.; Zhu, Q.; Zheng, X.; Zhang, K. A new antibacterial chromone derivative from mangrove-derived fungus *Penicillium aculeatum* (No. 9EB). *Nat. Prod. Res.* **2017**, *31*, 2593–2598. [CrossRef] [PubMed]

24. Chen, S.; Liu, Y.; Liu, Z.; Cai, R.; Lu, Y.; Huang, X.; She, Z. Isocoumarins and benzofurans from the mangrove endophytic fungus *Talaromyces amestolkiae* possess α-glucosidase inhibitory and antibacterial activities. *RSC Adv.* **2016**, *6*, 26412–26420. [CrossRef]

25. Chen, S.; He, L.; Dongni, C.; Cai, R.; Long, Y.; Lu, Y.; She, Z. Talaramide A, an unusual alkaloid from the mangrove endophytic fungus *Talaromyces* sp.(HZ-YX1) as inhibitor of mycobacterial PknG. *New J. Chem.* **2017**, *41*, 4273–4276. [CrossRef]

26. Frisvad, J.C.; Yilmaz, N.; Thrane, U.; Rasmussen, K.B.; Houbraken, J.; Samson, R.A. *Talaromyces atroroseus*, a new species efficiently producing industrially relevant red pigments. *PLoS ONE* **2013**, *8*, e84102. [CrossRef] [PubMed]

27. Li, H.; Huang, H.; Shao, C.; Huang, H.; Jiang, J.; Zhu, X.; Liu, Y.; Liu, L.; Lu, Y.; Li, M.; et al. Cytotoxic norsesquiterpene peroxides from the endophytic fungus *Talaromyces flavus* isolated from the mangrove plant *Sonneratia apetala*. *J. Nat. Prod.* **2011**, *74*, 1230–1235. [CrossRef] [PubMed]

28. Sridhar, K.R.; Mangalagangotri, M. Fungal diversity of Pichavaram mangroves, Southeast coast of India. *Nat. Sci.* **2009**, *7*, 67–75.

29. Liu, Y.; Xia, G.; Li, H.; Ma, L.; Ding, B.; Lu, Y.; He, L.; Xia, X.; She, Z. Vermistatin derivatives with α-glucosidase inhibitory activity from the mangrove endophytic fungus *Penicillium* sp. HN29-3B1. *Planta Med.* **2014**, *80*, 912–917. [CrossRef] [PubMed]

30. Xing, X.; Guo, S. Fungal endophyte communities in four Rhizophoraceae mangrove species on the south coast of China. *Ecol. Res.* **2011**, *26*, 403–409. [CrossRef]

31. Costa, I.P.; Maia, L.C.; Cavalcanti, M.A. Diversity of leaf endophytic fungi in mangrove plants of northeast Brazil. *Braz. J. Microbiol.* **2012**, *43*, 1165–1173. [CrossRef]

32. Lin, Z.; Zhu, T.; Fang, Y.; Gu, Q.; Zhu, W. Polyketides from *Penicillium* sp. JP-1, an endophytic fungus associated with the mangrove plant *Aegiceras corniculatum. Phytochemistry* **2008**, *69*, 1273–1278. [CrossRef] [PubMed]

33. Jin, P.F.; Zuo, W.J.; Guo, Z.K.; Mei, W.L.; Dai, H.F. Metabolites from the endophytic fungus *Penicillium* sp. FJ-1 of *Ceriops tagal. Acta Pharm. Sin.* **2013**, *48*, 1688–1691.

34. Zheng, C.; Chen, Y.; Jiang, L.L.; Shi, X.M. Antiproliferative metabolites from the endophytic fungus *Penicillium* sp. FJ-1 isolated from a mangrove *Avicennia marina. Phytochem. Lett.* **2014**, *10*, 272–275. [CrossRef]

35. Liu, J.F.; Chen, W.J.; Xin, B.R.; Lu, J. Metabolites of the endophytic fungus *Penicillium* sp. FJ-1 of *Acanthus ilicifolius. Nat. Prod. Commun.* **2014**, *9*, 799–801. [PubMed]

36. Liu, H.; Chen, S.; Liu, W.; Liu, Y.; Huang, X.; She, Z. Polyketides with immunosuppressive activities from mangrove endophytic fungus *Penicillium* sp. ZJ-SY2. *Mar. Drugs* **2016**, *14*, 217. [CrossRef] [PubMed]

37. Liu, F.; Li, Q.; Yang, H.; Cai, X.L.; Xia, X.K.; Chen, S.P.; Li, M.F.; She, Z.G.; Lin, Y.C. Structure elucidation of three diphenyl ether derivatives from the mangrove endophytic fungus SBE-14 from the South China Sea. *Magn. Reson. Chem.* **2009**, *47*, 453–455. [CrossRef] [PubMed]

38. Liu, F.; Cai, X.L.; Yang, H.; Xia, X.K.; Guo, Z.Y.; Yuan, J.; Li, M.F.; She, Z.G.; Lin, Y.C. The bioactive metabolites of the mangrove endophytic fungus *Talaromyces* sp. ZH-154 isolated from *Kandelia candel* (L.) Druce. *Planta Med.* **2010**, *76*, 185–189. [CrossRef] [PubMed]

39. Cai, R.; Chen, S.; Long, Y.; She, Z. Depsidones from *Talaromyces stipitatus* SK-4, an endophytic fungus of the mangrove plant *Acanthus ilicifolius. Phytochem. Lett.* **2017**, *20*, 196–199. [CrossRef]

40. Sreeta, K.; Dethoup, T.; Singburaudum, N.; Kijjoa, A. Antifungal activities of the crude extracts of endophytic fungi isolated from mangrove plants against phytopathogenic fungi in vitro. In *Proceedings of the 52nd Kasetsart University Annual Conference, Agricultural Sciences: Leading Thailand to World Class Standards, Kasetsart, Thailand, 4–7 February 2014*; Kasetsart University: Bangkok, Thailand, 2014; Volume 1, pp. 372–379.

41. Liu, Y.; Yang, Q.; Xia, G.; Huang, H.; Li, H.; Ma, L.; Lu, Y.; He, L.; Xia, X.; She, Z. Polyketides with α-glucosidase inhibitory activity from a mangrove endophytic fungus, *Penicillium* sp. HN29-3B1. *J. Nat. Prod.* **2015**, *78*, 1816–1822. [CrossRef] [PubMed]

42. Keller, N.P.; Turner, G.; Bennett, J.W. Fungal secondary metabolism—From biochemistry to genomics. *Nat. Rev. Microbiol.* **2005**, *3*, 937–947. [CrossRef] [PubMed]

43. Hussain, H.; Al-Sadi, A.M.; Schulz, B.; Steinert, M.; Khan, A.; Green, I.R.; Ahmed, I. A fruitful decade for fungal polyketides from 2007 to 2016: Antimicrobial activity, chemotaxonomy and chemodiversity. *Future Med. Chem.* **2017**, *9*, 1631–1648. [CrossRef] [PubMed]

44. Shukla, V.; Joshi, G.P.; Rawat, M.S.M. Lichens as a potential natural source of bioactive compounds: A review. *Phytochem. Rev.* **2010**, *9*, 303–314. [CrossRef]

45. Nicoletti, R.; Manzo, E.; Ciavatta, M.L. Occurence and bioactivities of funicone-related compounds. *Int. J. Mol. Sci.* **2009**, *10*, 1430–1444. [CrossRef] [PubMed]

46. Baroni, A.; De Luca, A.; De Filippis, A.; Petrazzuolo, M.; Manente, L.; Nicoletti, R.; Tufano, M.A.; Buommino, E. 3-*O*-methylfunicone, a metabolite of *Penicillium pinophilum*, inhibits proliferation of human melanoma cells by causing $G_2 + M$ arrest and inducing apoptosis. *Cell Prolif.* **2009**, *42*, 541–553. [CrossRef] [PubMed]

47. Buommino, E.; Paoletti, I.; De Filippis, A.; Nicoletti, R.; Ciavatta, M.L.; Menegozzo, S.; Menegozzo, M.; Tufano, M.A. 3-*O*-Methylfunicone, a metabolite produced by *Penicillium pinophilum*, modulates ERK1/2 activity, affecting cell motility of human mesothelioma cells. *Cell Prolif.* **2010**, *43*, 114–123. [CrossRef] [PubMed]

48. Buommino, E.; Tirino, V.; De Filippis, A.; Silvestri, F.; Nicoletti, R.; Ciavatta, M.L.; Pirozzi, G.; Tufano, M.A. 3-*O*-methylfunicone, from *Penicillium pinophilum*, is a selective inhibitor of breast cancer stem cells. *Cell Prolif.* **2011**, *44*, 401–409. [CrossRef] [PubMed]

49. Buommino, E.; De Filippis, A.; Nicoletti, R.; Menegozzo, M.; Menegozzo, S.; Ciavatta, M.L.; Rizzo, A.; Brancato, V.; Tufano, M.A.; Donnarumma, G. Cell-growth and migration inhibition of human mesothelioma cells induced by 3-*O*-methylfunicone from *Penicillium pinophilum* and cisplatin. *Investig. New Drugs* **2012**, *30*, 1343–1351. [CrossRef] [PubMed]

50. Nicoletti, R.; Scognamiglio, M.; Fiorentino, A. Structural and bioactive properties of 3-*O*-methylfunicone. *Mini Rev. Med. Chem.* **2014**, *14*, 1043–1047. [CrossRef]

51. Wu, C.; Zhao, Y.; Chen, R.; Liu, D.; Liu, M.; Proksch, P.; Guo, P.; Lin, W. Phenolic metabolites from mangrove-associated *Penicillium pinophilum* fungus with lipid-lowering effects. *RSC Adv.* **2016**, *6*, 21969–21978. [CrossRef]

52. Masters, K.S.; Bräse, S. Xanthones from fungi, lichens, and bacteria: The natural products and their synthesis. *Chem. Rev.* **2012**, *112*, 3717–3776. [CrossRef] [PubMed]

53. Negreiros de Carvalho, P.L.; de Oliveira Silva, E.; Chagas-Paula, D.A.; Hortolan Luiz, J.H.; Ikegaki, M. Importance and implications of the production of phenolic secondary metabolites by endophytic fungi: A mini-review. *Mini Rev. Med. Chem.* **2016**, *16*, 259–271. [CrossRef] [PubMed]

54. Shamsuzzaman, H.K. Bioactive benzofuran derivatives: A review. *Eur. J. Med. Chem.* **2015**, *97*, 483–504.

55. Saeed, A. Isocoumarins, miraculous natural products blessed with diverse pharmacological activities. *Eur. J. Med. Chem.* **2016**, *116*, 290–317. [CrossRef] [PubMed]

56. Mapari, S.A.; Thrane, U.; Meyer, A.S. Fungal polyketide azaphilone pigments as future natural food colorants? *Trends Biotechnol.* **2010**, *28*, 300–307. [CrossRef] [PubMed]

57. Osmanova, N.; Schultze, W.; Ayoub, N. Azaphilones: A class of fungal metabolites with diverse biological activities. *Phytochem. Rev.* **2010**, *9*, 315–342. [CrossRef]

58. Gao, J.M.; Yang, S.X.; Qin, J.C. Azaphilones: Chemistry and biology. *Chem. Rev.* **2013**, *113*, 4755–4811. [CrossRef] [PubMed]

59. Barton, D.H.R.; Sutherland, J.K. The nonadrides. Part I. Introduction and general survey. *J. Chem. Soc.* **1965**, 1769–1772. [CrossRef]

60. Bloomer, J.L.; Moppett, C.E.; Sutherland, J.K. The nonadrides. Part V. Biosynthesis of glauconic acid. *J. Chem. Soc. C Org.* **1968**, 588–591. [CrossRef]

61. Cox, R.E.; Holker, J.S. Biosynthesis of glauconic acid from [2, 3–13 C] succinate. *J. Chem. Soc. Chem. Commun.* **1976**, *15*, 583–584. [CrossRef]

62. Zhang, Y.; Han, T.; Ming, Q.; Wu, L.; Rahman, K.; Qin, L. Alkaloids produced by endophytic fungi: A review. *Nat. Prod. Commun.* **2012**, *7*, 963–968. [PubMed]

63. West, R.R.; Van Ness, J.; Varming, A.M.; Rassing, B.; Biggs, S.; Gasper, S.; Mckernan, P.A.; Piggott, J. ZG-1494α, a novel platelet-activating factor acetyltransferase inhibitor from *Penicilium rubrum*, isolation, structure elucidation and biological activity. *J. Antibiot.* **1996**, *49*, 967–973. [CrossRef] [PubMed]

64. Higginbotham, S.J.; Arnold, A.E.; Ibañez, A.; Spadafora, C.; Coley, P.D.; Kursar, T.A. Bioactivity of fungal endophytes as a function of endophyte taxonomy and the taxonomy and distribution of their host plants. *PLoS ONE* **2013**, *8*, e73192. [CrossRef] [PubMed]

65. El-Elimat, T.; Figueroa, M.; Ehrmann, B.M.; Cech, N.B.; Pearce, C.J.; Oberlies, N.H. High-resolution MS, MS/MS, and UV database of fungal secondary metabolites as a dereplication protocol for bioactive natural products. *J. Nat. Prod.* **2013**, *76*, 1709–1716. [CrossRef] [PubMed]

66. Nielsen, K.F.; Larsen, T.O. The importance of mass spectrometric dereplication in fungal secondary metabolite analysis. *Front. Microbiol.* **2015**, *6*, 71. [CrossRef] [PubMed]

67. Jones, M.B.; Nierman, W.C.; Shan, Y.; Frank, B.C.; Spoering, A.; Ling, L.; Peoples, A.; Zullo, A.; Lewis, K.; Nelson, K.E. Reducing the bottleneck in discovery of novel antibiotics. *Microb. Ecol.* **2017**, *73*, 658–667. [CrossRef] [PubMed]

marine drugs

MDPI

Communication

Two New Terpenoids from *Talaromyces purpurogenus*

Wenjing Wang, Xiao Wan, Junjun Liu, Jianping Wang, Hucheng Zhu, Chunmei Chen * and Yonghui Zhang *

Hubei Key Laboratory of Natural Medicinal Chemistry and Resource Evaluation, Tongji Medical College, Huazhong University of Science and Technology, Wuhan 430030, China; wangwj0122@163.com (W.W.); marina.wanx@gmail.com (X.W.); junjun.liu@hust.edu.cn (J.L.); jpwang1001@163.com (J.W.); zhuhucheng@hust.edu.cn (H.Z.)

* Correspondence: chenchunmei@hust.edu.cn (C.C.); zhangyh@mails.tjmu.edu.cn (Y.Z.);
  Tel.: +86-27-8369-2892 (C.C.); +86-27-8369-2892 (Y.Z.)

Received: 30 March 2018; Accepted: 27 April 2018; Published: 2 May 2018

Abstract: A new sesquiterpenoid 9,10-diolhinokiic acid (**1**) and a new diterpenoid roussoellol C (**2**), together with 4 known compounds, were isolated from the extracts of laboratory cultures of marine-derived fungus *Talaromyces purpurogenus*. 9,10-diolhinokiic acid is the first thujopsene-type sesquiterpenoid containing a 9,10-diol moiety, and roussoellol C possesses a novel tetracyclic fusicoccane framework with an unexpected hydroxyl at C-4. These new structures were confirmed by spectroscopic data, chemical method, NMR data calculations and electronic circular dichroism (ECD) calculations. The selected compounds were evaluated for cytotoxicities against five human cancer cell lines, including SW480, HL-60, A549, MCF-7, and SMMC-7721 and the IC_{50} values of compound **2** against MCF-7 and **3** against HL-60 cells were 6.5 and 7.9 μM, respectively.

Keywords: sesquiterpenoid; diterpenoid; *Talaromyces purpurogenus*; NMR data calculations; ECD calculations; cytotoxicities

1. Introduction

Over the past forty years, more than 60% small molecule new drugs have been directly or indirectly derived from natural product source, which demonstrates that natural products continue to play a significant role in drug discovery and development process [1]. Fungi-derived natural products are rich sources of medicines due to their diverse chemical structures and bioactivities. For example, lovastatin, penicillin, echinocandin B, and cyclosporine A have been clinically used as effective medicines, illustrating the significance of fungi-derived metabolites in drug discovery [2].

The fungus *Talaromyces purpurogenus*, previously known as *Penicillium purpurogenum* [3], is widely distributed in terrestrial plants, soil, and marine habitats, and has been reported to produce various secondary metabolites such as meroterpenoids [4,5], polyketides [6–10], lipopeptides [10], and sterols [11]. Meanwhile, the impressive structurally diverse metabolites from this fungus exhibit extensive bioactivities including anti-inflammatory [11], anti-influenza virus [7], insecticidal [4], antitumor [9], and antifungal activities [8]. In our screening of extracts of several fungi for their cytotoxic activities, EtOAc extract of *T. purpurogenus*, isolated from a mud sample, showed significant cytotoxic activity in vitro. A chemical investigation of the fungus *T. purpurogenus* resulted in the isolation of two new secondary metabolites 9,10-diolhinokiic acid (**1**) and roussoellol C (**2**), and four known compounds including dankasterone (**3**) [12,13], cyclotryprostatin E (**4**) [14], 6-methoxyspirotryprostatin B (**5**) [15], and (3*S*,12a*S*)-3-methyl-2,3,6,7,12,12a-hexahydropyrazino[1',2':1,6]pyrido[3,4-b]indole-1,4-dion (**6**) (Figure 1) [16]. Details of isolation, structural elucidation, and cytotoxic activities are presented here.

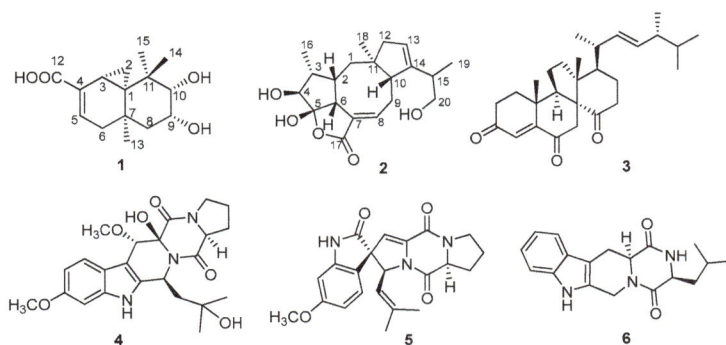

Figure 1. The structures of compounds **1–6**.

2. Results and Discussion

2.1. Chemical Identification of Isolated Terpenoids

Compound **1** was isolated as pale yellow oil. The molecular formula of $C_{15}H_{22}O_4$, containing 5 degrees of unsaturation, was deduced from its HRESIMS spectrum (m/z 265.1450 [M − H]$^-$; calcd. for $C_{15}H_{21}O_4$, 265.1440). The IR spectrum gave a hydroxyl absorption band at 3432 cm^{-1} and unsaturated carboxyl and double bond absorbances at 1696 and 1647 cm^{-1}, respectively. The ^{1}H NMR data (Table 1) of **1** showed the existence of one olefinic proton at δ_H 6.56 (1H, dd, J = 7.0, 1.8 Hz); two oxygenated protons at δ_H 3.23 (1H, d, J = 3.8 Hz) and 4.03 (1H, ddd, J = 3.8, 3.7, 3.1 Hz); three methyl signals at δ_H 1.44, 1.31, and 0.78. The ^{13}C NMR and DEPT data (Table 1) showed only 13 carbon resonances, comprising one olefinic carbon [δ_C 133.7 (C-5)], three methyls [δ_C 30.6 (C-13), 25.9 (C-14), and 23.5 (C-15)], three methylenes [δ_C 11.3 (C-2), 42.8 (C-6), and 41.5 (C-8)], three methines including two oxygenated carbons [δ_C 17.9 (C-3), 72.8 (C-9), and 78.6 (C-10)], as well as three quaternary carbons [δ_C 35.7 (C-1), 31.9 (C-7), and 39.9 (C-11)]. The missing carbons in the ^{13}C NMR including a carboxyl (δ_C 171.2) and an olefinic carbon (δ_C 134.3) were revealed by cross-peaks in the HMBC spectrum. These data indicated compound **1** to be a sesquiterpenoid.

Exhaustive analyses of the 2D NMR spectra of **1** revealed some similarity to (+)-thujopsene [17], a sesquiterpenoid derivative isolated from the liverwort *Marchantia polymorpha*. However, two oxygenated methines (δ_H 4.03, δ_C 72.8 and δ_H 3.23, δ_C 78.6) of **1** replaced the methylenes of (+)-thujopsene and one methyl was oxidized to a carboxyl (δ_C 171.2) group. The presence of a 9,10-diol moiety was demonstrated by the ^{1}H–^{1}H COSY cross-peak of H-9 and H-10, and HMBC correlations from H-9 to C-7 and C-11 and from H-10 to C-8 and C-11. The carboxyl was located at C-4, which was substantiated by the HMBC cross-peaks from H-3 and H-5 to C-12 (Figure 2). Therefore, the planar structure of **1** was identified as a 9, 10-diolhinokiic acid [18]. In the NOESY experiment (Figure 2), the correlations of H-2/Me-15 and H-2/Me-13 suggested that these groups were co-facial and assigned as α-oriented, while, the interactions of H-3/Me-14 and Me-14/H-10 indicated that they were on the opposite face of the ring system and β-oriented. The hydroxyls at C-9 and C-10 were on the same side according to the coupling constant between H-9 and H-10 (J = 3.8 Hz). Thus, the relative configuration of **1** was determined. The absolute stereochemistry of the 9,10-diol moiety in **1** was verified by observing the induced electronic circular dichroism (IECD) spectrum after the addition of dimolybdenum tetraacetate in anhydrous DMSO [19,20]. The obvious negative Cotton effect at 310 nm in the IECD spectrum (Figure 3) permitted the 9R,10S configuration assignment of **1**. Combining with the relative configuration, the absolute stereochemistry of **1** was elucidated as 1R,3R,7R,9R,10S (Figure 1).

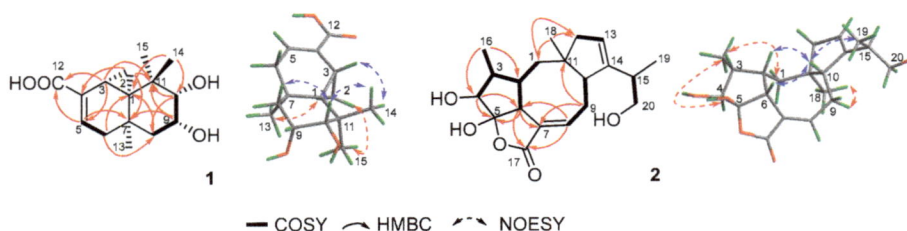

Figure 2. Key 2D correlations of compounds **1** and **2**.

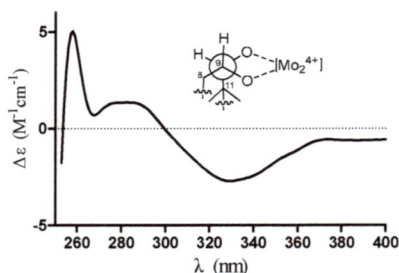

Figure 3. Conformation of the Mo$_2$$^{4+}$ complex of compound **1** and its IECD spectrum in DMSO.

Compound **2** was isolated as colorless oil that gave a [M + Na]$^+$ ion peak in the HRESIMS spectrum at *m/z* 371.1839 Δm (calcd. for C$_{20}$H$_{28}$O$_5$Na, 371.1834) appropriate for a molecular formula of C$_{20}$H$_{28}$O$_5$, corresponding to 7 degrees of unsaturation. The IR spectrum showed a hydroxyl (3430 cm^{-1}) and an ester or lactone carbonyl (1742 cm^{-1}). The ^{1}H NMR data (Table 1) of **2** revealed two olefinic protons at δ_H 7.04 (1H, m) and 5.31 (1H, s); three oxygenated protons at δ_H 3.81 (1H, d, *J* = 4.4 Hz), 3.66 (1H, dd, *J* = 10.6, 5.3 Hz), and 3.37 (1H, dd, *J* = 10.6, 7.9 Hz); three methyl signals at δ_H 0.84 (3H, s), 0.93 (3H, d, *J* = 7.4 Hz), and 1.04 (3H, d, *J* = 6.8 Hz). The ^{13}C NMR and DEPT spectra (Table 1) displayed resonances for 20 carbon signals categorized as one carbonyl carbon [δ_C 173.0 (C-17)], four olefinic carbons [δ_C 130.5 (C-7), 141.6 (C-8), 122.1 (C-13), and 148.4 (C-14)], one sp^3 quaternary carbon [δ_C 46.9 (C-11)], one hemiketal carbon [δ_C 114.1 (C-1)], six methines including one oxygenated carbon [δ_C 41.2 (C-2), 40.8 (C-3), 80.0 (C-4), 52.8 (C-6), 47.9 (C-10), and 36.7 (C-15)], four methylenes including one oxygenated carbon [δ_C 37.6 (C-1), 26.8 (C-9), 45.5 (C-12), and 67.0 (C-20)], and three methyls [δ_C 11.0 (C-16), 24.9 (C-18), and 17.7 (C-19)]. Consideration of these data and analyses of the ^{1}H–^{1}H COSY and HMBC spectra (Figure 2) of **2** suggested existence of tetracyclic fusicoccane framework which was similar with that of roussoellol B [21]. Further analyses of the 2D NMR spectra indicated that the methylene at C-4 in roussoellos B was oxygenated (δ_C 80.0), which was confirmed by ^{1}H–^{1}H COSY cross-peak of H-3 and H-4 and the HBMC correlations from H-4 to C-3, C-5, C-6, and C-16. In addition, the $\Delta^{10,14}$-double bond in roussoello B shifted to C-13 and C-14 which was evidenced by the HBMC correlations from H-13 to C-10, C-11, and C-12. Moreover, HMBC correlations from H-15 and H-19 to the oxygenated methylene carbon at δ_C 67.0 (C-20) indicated the presence of a hydroxymethyl functionality. Hence, the planar structure of **2** was established as shown.

In the NOESY experiment (Figure 2), the NOESY correlations of H-2/H-10, H-6/H-10, and H-10/Me-19 indicated the β-orientation for these protons. Meanwhile, correlations of H-1/Me-16, Me-16/H-4, and Me-18/H-9 suggested that H-4, Me-16, and Me-18 were on the opposite side and α-oriented. Even though the NOESY correlation of H-10/Me-19 were observed, but the configuration of C-15 could not be determined by the NOESY experiment due to the freely rotation of the bond between C-14 and C-15. In order to determine the relative configuration of 5-OH, the theoretical calculation of ^{13}C NMR chemical shifts of epimers **2a** and **2b** (Figure 4) were performed to semiempirical PM3 quantum mechanical geometry optimizations using Gaussian09 at the B3LYP/6-31G* level [22]. The experimental shifts were plotted against the calculated shifts, and least-squares fit lines was confirmed. The calculated shifts for **2a** and **2b** were corrected by the slope and intercept to get the corrected ^{13}C shifts (Table 1), and the differences between the corrected and experimental ^{13}C NMR chemical shifts were analyzed [23,24]. The result showed that the correlation coefficient R^2 of **2a** (0.9966) was higher than that of **2b** (0.9926) (Figures S21 and S22). Meanwhile, the MAE (mean absolute error) and MD (maximum deviation) of **2a** (MAE = 2.28, MD = 6.5) were obviously lower than that of **2b** (MAE = 2.74, MD = 14.3), suggesting that **2a** was more consistent with the experimental values (Figure 4). What's more, all of the reported fusicoccanes or ophiobolins with 5/8/5/5 ring system possess a *cis*-fused A/D ring [21,25–28], and the 5-OH of **2** was finally assigned a β-orientation as **2a**.

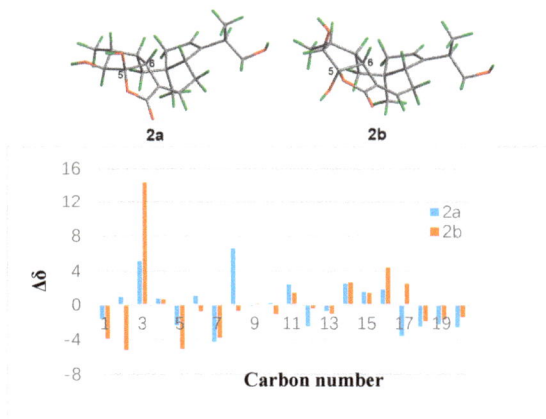

Figure 4. Structures and differences in ppm between calculated and experimental ^{13}C NMR shifts for **2a** and **2b**.

To determine the absolute configuration of compound **2**, the electronic circular dichroism (ECD) calculation was performed. The experimental and simulated spectra generated by BALLOON [29,30] were performed to semiempirical PM3 quantum mechanical geometry optimizations using the Gaussian 09 program (Figures S2 and S3, Supplementary Materials) [31]. The ECD spectrum of each conformer was calculated using the TDDFT methodology at B3LYP/6-311++G(d,p)//B3LYP/6-31G(d) level. Comparison of the experimental and calculated spectra of **2** showed more agreement (Figure 5) for the **2a** configuration. The experimental ECD is consistent with the calculated ECD of **2** (Figure 5), indicating a (2*S*,3*R*,4*S*,5*S*,6*R*,10*R*,11*S*)-configuration. Therefore, the structure of **2**, namely, roussoellol C, was deduced as shown.

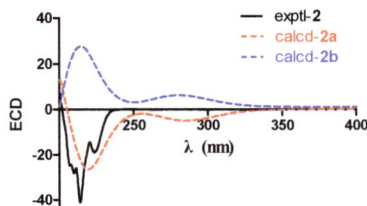

Figure 5. Experimental and calculated ECD spectra of **2**.

Table 1. ^{1}H (400 MHz) and ^{13}C NMR (100 MHz) Data for Compound **1** and **2** (CD$_3$OD) and DFT Calculation of ^{13}C NMR for **2a** and **2b**.

Position	1		2		2a		2b	
	δ_H (*J* in Hz)	δ_C	δ_H (*J* in Hz)	δ_C	δ_C (calcd.)	δ_C (cor)	δ_C (calcd.)	δ_C (cor)
1	-	35.7	1.60, d, 13.0 1.40, dd, 14.6, 13.0	37.6	39.2	35.9	36.2	33.7
2	0.85, dd, 9.1, 5.0 0.76, d, 5.1	11.3	2.70, m	41.2	45.7	42.1	38.6	36.0
3	2.11, dd, 9.0, 5.1	17.9	2.45, m	40.8	49.6	45.9	58.7	55.1
4	-	134.3	3.81, d, 4.4	80.0	86.0	80.7	85.4	80.6
5	6.56, d, 4.6	133.7	-	114.1	118.4	111.7	115.2	109.0
6	1.88, dd, 18.3, 2.6 1.78, dd, 18.3, 7.0	42.8	3.42, br d, 9.4	52.8	57.9	53.8	55.4	52.0
7	-	31.9	-	130.5	133.5	126.2	133.7	126.7
8	1.57, dd, 14.4, 3.1 1.50, dd, 14.4, 3.7	41.5	7.04, m	141.6	156.4	148.1	148.6	140.9
9	4.03, ddd, 3.8, 3.7, 3.1	72.8	2.65, overlap 2.36, m	28.6	31.4	28.5	31.0	28.7
10	3.23, d, 3.8	78.6	3.19, dd, 13.6, 2.8	47.9	51.9	48.1	50.0	46.8
11	-	39.9	-	46.9	53.2	49.3	51.6	48.3
12	-	171.2	2.28, dd, 15.1, 4.2 1.76, dd, 15.1, 2.6	45.5	46.6	43.0	48.2	45.9
13	1.44, s	30.6	5.31, br s	122.1	128.4	121.3	127.8	121.0
14	0.78, s	25.9	-	148.4	159.3	150.9	159.3	151.1
15	1.31, s	23.5	2.22, ddq, 7.9, 5.3, 7.4	36.7	41.6	38.2	40.8	38.1
16			0.93, d, 7.4	11.0	15.0	12.8	17.0	15.4
17				173.0	178.6	169.4	184.9	175.5
18			0.84, s	24.9	25.0	22.4	25.0	23.0
19			1.04, d, 6.8	17.7	17.7	15.4	17.5	15.8
20			3.66, dd, 10.6, 5.3 3.37, dd, 10.6, 7.9	67.0	68.9	64.4	69.6	65.5

2.2. Cytotoxic Activities of Selected Compounds

The growth inhibitory effects of the selected compounds (**1–3**) against human colonic carcinoma cell line (SW480), human promyelocytic leukemia cells (HL-60), human non-small-cell lung cancer cells (A549), breast adenocarcinoma cell line (MCF-7), and human hepatocellular carcinoma cell line (SMMC-7721) were assayed by using MTT method [32], with adriamycin as the positive control. Compounds **1–3** exhibited moderate antiproliferative activities against these cells with IC_{50} values ranging from 6.5 to 35.7 μM (Table 2). Normally, cytotoxic natural products display better activities against HL-60 than any other cancer cell lines because HL-60 cells are much sensitive in the assay. However, it is interesting that compound **2** showed significant selectivity toward MCF-7 cells with an IC_{50} value of 6.5 μM but with an IC_{50} value of 10.9 μM against HL-60. Although an IC_{50} value of 6.5 μM does not indicate strong potency, the selectivity of **2** against MCF-7 still makes it a promising lead compound for further studies.

Table 2. Cytotoxicities against Tumor Cells for **1–3** (IC_{50}, μM).

	1	2	3	Adriamycin
SW480	>40	23.6	14.2	1.2
HL-60	12.6	10.9	7.9	0.05
A549	35.7	25.8	21.3	0.10
MCF-7	>40	6.5	23.8	0.80
SMMC-7721	>40	>40	>40	0.2

3. Materials and Methods

3.1. General Experimental Procedures

Optical rotations were measured on a Rudolph Autopol IV automatic polarimeter with a 0.7 mL cell (Rudolph Research Analytical, Hackettstown, NJ, USA). UV spectra were recorded with a PerkinElmer Lambda 35 spectrophotometer (PerkinElmer, Inc., Fremont, CA, USA). IR spectra were measured on a Bruker Vertex 70 FT-IR spectrophotometer (Bruker, Karlsruhe, Germany). ECD data were obtained with a JASCO-810 instrument (JASCO Co., Ltd., Tokyo, Japan). HRESIMS data were recorded on a Thermo Fisher LTQ XL LC/MS (Thermo Fisher, Palo Alto, CA, USA). 1D and 2D NMR spectra were measured with a Bruker AM-400 NMR spectrometer at 25 °C (Bruker, Karlsruhe, Germany), the NOESY mixing time was 100 ms. Compounds were purified by an Agilent 1220 HPLC system semi-preparative HPLC (Agilent Technologies Inc., Santa Clara, CA, USA) equipped with a UV detector (Agilent Technologies Inc.). Column chromatography was performed on silica gel (100–200 and 200–300 mesh, Qingdao Marine Chemical Inc., Qingdao, China), Sephadex LH-20 (Pharmacia Biotech AB, Uppsala, Sweden) and ODS (50 μm, YMC, Kyoto, Japan).

3.2. Fungal Material

The fungus PP-414 was isolated from a mud sample collected on the coastal beach in Qinghuangdao County, Hebei Province, China. The mud sample (5 g) was suspended in 50 mL sterile water with a concentration at 10^{-1} g/mL and then every 0.5 mL mutterlauge was respectively diluted to 10^{-2}, 10^{-3}, 10^{-4} g/mL with sterile water. Each sample was coated individually on potato dextrose agar (PDA) medium contained chloramphenicol, and incubated at 28 °C to get single colonies by routine microbiological methods. The internal transcribed spacer (ITS) region was amplified by PCR using primers ITS1 (5′-TCCGTAGGTGAACCTGCGG-3′) and ITS4 (5′-TCCTCCGCTTATTGATATGC-3′), then submitted to GenBank and identified as *Talaromyces purpurogenus* by ITS sequence homology (99% similarity with *Talaromyces purpurogenus* strain Q2, accession no. KX432212.1 (max score 974, query cover 96%, *e* value 0.0)) and physiological characteristics with accession no. MH120320. The voucher sample, PP-414, has been preserved in the culture collection center of Tongji Medical College, Huazhong University of Science and Technology (Wuhan, China).

3.3. Fermentation and Isolation

The fungus PP-414 was incubated on potato dextrose agar (PDA) at 28 °C for 7 days, the agar cultures were cut into small pieces (approximately $0.5 \times 0.5 \times 0.5$ cm^3) and then inoculated into 100×1 L Erlenmeyer flasks which containing 250 g rice and 250 mL distilled water. After incubating at 28 °C for 28 days, the solid rice medium was distilled with CH_3CH_2OH and then extracted three times with EtOAc. The EtOAc extract (80 g) was chromatographed on silica gel chromatography column (CC, 80–120 mesh) eluting with petroleum ether/EtOAc (100:0–0:1, *v/v*) to afford five fractions (Fr. A–Fr. E). Fr. C (4.5 g) was further separated by Sephadex LH-20 (CH_2Cl_2/MeOH 1:1) and silica gel CC (200–300 mesh) eluting with CH_2Cl_2/MeOH (200:1–20:1, *v/v*) to obtain four fractions (C1-C5), Fr. C2 was further purified by semi-preparative HPLC (MeCN-H$_2$O, 85:15, *v/v*) to yield **3** (9.0 mg, t_R = 52.5 min). Fr. C3 was further purified by semi-preparative HPLC (MeCN-H$_2$O, 45:55, *v/v*) to obtain **6** (6.0 mg, t_R = 35.2 min). Fr. C4 was further separated by Sephadex LH-20 (MeOH) and semi-preparative HPLC (MeOH-H$_2$O, 65:35, *v/v*) to yield **4** (4.5 mg, t_R = 20.0 min) and **5** (5.0 mg, t_R = 46.0 min).

Fr. D (8.5 g) was further separated by reversed-phase MPLC (MeOH/H$_2$O, 10:90–100:0) to obtain seven fractions (D1–D7), Fr. D6 was purified by Sephadex LH-20 (CH_2Cl_2/MeOH 1:1) and silica gel CC (200–300 mesh) eluting with CH_2Cl_2/MeOH (200:1–10:1, *v/v*) to obtain four fractions (D6.1–D6.4), Fr. D6.3 was further purified by semi-preparative HPLC (MeOH-H$_2$O, 65:35, *v/v*) to yield **1** (2.5 mg, t_R = 16.0 min). Fr. D6.4 was further separated by semi-preparative HPLC (MeOH-H$_2$O, 55:45, *v/v*) to yield **2** (1.1 mg, t_R = 27.5 min).

Compound 1: pale yellow oil (MeOH), $[\alpha]_D^{25}$ + 43.0 (*c* 0.1, MeOH); UV (MeOH) λ_{max} (log ε) 203 (3.71), 235 (3.51) nm; CD (MeOH) λ_{max} ($\Delta\varepsilon$) 203 (−1.37), 233 (2.17), 263 (0.54), 282 (0.72), 335 (−0.13) nm; IR ν_{max} 3432, 2923, 1696, 1647, 1382, 1246, 1062 cm^{-1}; ^{1}H and ^{13}C NMR data, see Table 1; HRESIMS *m/z* 265.1450 [M − H]$^-$ (calcd. for $C_{15}H_{21}O_4$, 265.1440).

Compound 2: colorless oil (MeOH), $[\alpha]_D^{25}$ − 4.0 (*c* 0.05, MeOH); UV (MeOH) λ_{max} (log ε) 202 (3.87), 212 (3.81), 225 (3.84) nm; CD (MeOH) λ_{max} ($\Delta\varepsilon$) 212 (−7.28), 232 (−2.44), 243 (−3.34) nm; IR ν_{max} 3430, 2928, 1742, 1631, 1384, 1218, 1027 cm^{-1}; ^{1}H and ^{13}C NMR data, see Table 1; HRESIMS *m/z* 371.1839 [M + Na]$^+$ (calcd. for $C_{20}H_{28}O_5Na$, 371.1834).

3.4. NMR Calculation

The ^{13}C NMR chemical shifts of each conformer were calculated at the B3LYP/6-311++G(d,p)// B3LYP/6-31G(d) level by the IEFPCM solvation model implemented using Gaussian 09 program with MeOH as solvent, which were then combined using Boltzmann weighting according to their population contributions. The detailed methods were the same as previously described [33].

3.5. ECD Calculation

The electronic circular dichroism (ECD) spectra of each conformer were calculated by the TDDFT methodology with MeOH as solvent. The detailed methods were the same as previously described [32] The ECD spectra of each conformer were simulated using a Gaussian function with a bandwidth σ of 0.4 eV. The spectra were combined after Boltzmann weighting according to their population contributions and UV correction was applied.

3.6. Cytotoxicity against Cancer Cell Lines

Cytotoxicity of the selected compounds against the five cancer cell lines (SW480, HL-60, A549, MCF-7, and SMMC-7721) was evaluated by the MTT method with adriamycin as positive control. All cells were cultured in RPMI.1640 medium contained 10% fetal bovine serum, 2 mM

L-glutamine, 100 U/mL penicillin, and 100 µg/mL streptomycin at 37 °C in a humidified atmosphere with 5% CO_2. Tumor cells were seeded in 96-well microtiter plates at 5000 cells/wel, and the test compounds at concentrations ranging from 1.56 to 50 µM were added to the wells 12 h later. After incubation for 48 h, the metabolic conversion of 20 µL of MTT (5 mg/mL) 3-(4,5-dimethylthiazol-2-yl)-2,5-diphenyltetrazolium bromide was added and the incubation was continued for 4 h at 37 °C. The medium was exchanged with the medium containing 100 µL triplex solution of 10% SDS, 5% isopropyl alcohol and 12 mM HCl and then cultured 12–20 h at 37 °C. The results were obtained using a microplate spectrophotometer plate reader at 570 nm and the value of inhibition was calculated by formula: % inhibition = [($OD_{control}$ − $OD_{treated}$)/$OD_{control}$] × 100%. Selected compounds were tested at five concentrations (50, 25, 12.5, 6.25, 3.12 and 1.56 µM) in 100% DMSO with a final concentration of DMSO was 0.5% (*v/v*) in each well. The IC_{50} values were calculated by the means ± SEM calculating by GraphPad Prism 5.

4. Conclusions

In conclusion, we have reported six metabolites, including two new structures from the culture extract of *T. purpurogenus*. Among them, 9,10-diolhinokiic acid (**1**) is the first reported thujopsene-type sesquiterpenoid containing a 9,10-diol moiety and roussoellol C (**2**) possesses a novel tetracyclic fusicoccane diterpenoid with an unexpected hydroxyl at C-4. This study further enriched the structure diversity of secondary metabolites of this species. Additionally, compound **2** showed significant selectivity aginst MF-7 with IC_{50} values of 6.5 µM, which makes it a promising lead compound for further studies.

Supplementary Materials: The following are available online at http://www.mdpi.com/1660-3397/16/5/150/s1, HR-ESI-MS, IR, UV and NMR spectra of new compounds **1** and **2**, as well as computational data for compound **2**.

Author Contributions: W.W. conducted the main experiments, data analyzes, and wrote the manuscript, X.W. and J.L. were responsible for the NMR and ECD calculations, J.W. conducted the fermentation of fungi, H.Z. analyzed the spectroscopic data, C.C. and Y.Z. designed the experiments and commented on the manuscript.

Funding: This work was financially supported by the Program for Changjiang Scholars of Ministry of Education of the People's Republic of China (No. T2016088); National natural Science Foundation for Distinguished Young Scholars (No. 81725021); Innovative Research Groups of the National Natural Science Foundation of China (81721005); the Academic Frontier Youth Team of HUST; the Integrated Innovative Team for Major Human Diseases Program of Tongji Medical College (HUST).

Acknowledgments: We thank the Analytical and Testing Center at Huazhong University of Science and Technology for assistance in testing of ECD, UV and IR analyses. We thank Chong Dai and Ying Tang for assistance in isolated experiment.

Conflicts of Interest: The authors declare no conflict of iterest.

References

1. Newman, D.J.; Cragg, G.M. Natural products as sources of new drugs from 1981 to 2014. *J. Nat. Prod.* **2016**, *79*, 629–661. [CrossRef] [PubMed]

2. Evidente, A.; Kornienko, A.; Cimmino, A.; Andolfi, A.; Lefrance, F.; Mathieu, V.; Kiss, R. Fungal metabolites with anticancer activity. *Nat. Prod. Rep.* **2014**, *31*, 617–627. [CrossRef] [PubMed]

3. Samson, R.A.; Yilmaz, N.; Houbraken, J.; Spierenburg, H.; Seifert, K.A.; Peterson, S.W.; Varga, J.; Frisvad, J.C. Phylogeny and nomenclature of the genus talaromyces and taxa accommodated in penicillium subgenus biverticillium. *Stud. Mycol.* **2011**, *70*, 159–183. [CrossRef] [PubMed]

4. Centko, R.M.; Williams, D.E.; Patrick, B.O.; Akhtar, Y.; Garcia Chavez, M.A.; Wang, Y.A.; Isman, M.B.; de Silva, E.D.; Andersen, R.J. Dhilirolides E–N, meroterpenoids produced in culture by the fungus *Penicillium purpurogenum* collected in Sri Lanka: Structure elucidation, stable isotope feeding studies, and insecticidal activity. *J. Org. Chem.* **2014**, *79*, 3327–3335. [CrossRef] [PubMed]

5. Silva, E.D.; Williams, D.E.; Jayanetti, D.R.; Centko, R.M.; Patrick, B.O.; Wijesundera, R.L.; Andersen, R.J. Dhilirolides A–D, meroterpenoids produced in culture by the fruit-infecting fungus *Penicillium purpurogenum* collected in Sri Lanka. *Org. Lett.* **2011**, *13*, 1174–1177. [CrossRef] [PubMed]

6. Sun, J.; Zhu, Z.X.; Song, Y.L.; Ren, Y.; Dong, D.; Zheng, J.; Liu, T.; Zhao, Y.F.; Tu, P.F.; Li, J. Anti-neuroinflammatory constituents from the fungus *Penicillium purpurogenum* MHZ 111. *Nat. Prod. Res.* **2017**, *31*, 562–567. [CrossRef] [PubMed]

7. Wang, H.; Wang, Y.; Wang, W.; Fu, P.; Liu, P.; Zhu, W. Anti-influenza virus polyketides from the acid-tolerant fungus *Penicillium purpurogenum* JS03-21. *J. Nat. Prod.* **2011**, *74*, 2014–2018. [CrossRef] [PubMed]

8. Li, H.; Wei, J.; Pan, S.Y.; Gao, J.M.; Tian, J.M. Antifungal, phytotoxic and toxic metabolites produced by *Penicillium purpurogenum*. *Nat. Prod. Res.* **2014**, *28*, 2358–2361. [CrossRef] [PubMed]

9. Xia, M.W.; Cui, C.B.; Li, C.W.; Wu, C.J.; Peng, J.X.; Li, D.H. Rare chromones from a fungal mutant of the marine-derived *Penicillium purpurogenum* G59. *Mar. Drugs* **2015**, *13*, 5219–5236. [CrossRef] [PubMed]

10. Wu, C.J.; Li, C.W.; Cui, C.B. Seven new and two known lipopeptides as well as five known polyketides: The activated production of silent metabolites in a marine-derived fungus by chemical mutagenesis strategy using diethyl sulphate. *Mar. Drugs* **2014**, *12*, 1815–1838. [CrossRef] [PubMed]

11. Xue, J.; Wu, P.; Xu, L.; Wei, X. Penicillitone, a potent in vitro anti-inflammatory and cytotoxic rearranged sterol with an unusual tetracycle core produced by *Penicillium purpurogenum*. *Org. Lett.* **2014**, *16*, 1518–1521. [CrossRef] [PubMed]

12. Amagata, T.; Doi, M.; Tohgo, M.; Minoura, K.; Numata, A. Dankasterone, a new class of cytotoxic steroids produced by a *Gymnascella* species from a marine sponge. *Chem. Commun.* **1999**, *30*, 1321–1322. [CrossRef]

13. Amagata, T.; Tanaka, M.; Yamada, T.; Doi, M.; Minoura, K.; Ohishi, H.; Yamori, T.; Numata, A. Variation in cytostatic constituents of a sponge-derived *gymnascella dankaliensis* by manipulating the carbon source. *J. Nat. Prod.* **2007**, *70*, 1731–1740. [CrossRef] [PubMed]

14. He, F.; Sun, Y.L.; Liu, K.S.; Zhang, X.Y.; Qian, P.Y.; Wang, Y.F.; Qi, S.H. Indole alkaloids from marine-derived fungus *Aspergillus sydowii* SCSIO 00305. *J. Antibiot.* **2012**, *65*, 109–111. [CrossRef] [PubMed]

15. Zhang, M.; Wang, W.L.; Fang, Y.C.; Zhu, T.J.; Gu, Q.Q.; Zhu, W.M. Cytotoxic alkaloids and antibiotic nordammarane triterpenoids from the marine-derived fungus *Aspergillus sydowi*. *J. Nat. Prod.* **2008**, *71*, 985–989. [CrossRef] [PubMed]

16. Wu, G.; Liu, J.; Bi, L.; Zhao, M.; Wang, C.; Baudy-Floc'h, M.; Ju, J.; Peng, S. Toward breast cancer resistance protein (BCRP) inhibitors: Design, synthesis of a series of new simplified fumitremorgin C analogues. *Tetrahedron* **2007**, *63*, 5510–5528. [CrossRef]

17. Matsuo, A.; Nakayama, N.; Nakayama, M. Enantiomeric type sesquiterpenoids of the liverwort *Marchantia polymorpha*. *Phytochemistry* **1985**, *24*, 777–781. [CrossRef]

18. Norin, T.; Jakobsen, H.J.; Larsen, E.H.; Forsén, S.; Meisingseth, E. The chemistry of the natural order Cupressales. 49. The configuration of thujopsene. *Acta Chem. Scand.* **1963**, *17*, 738–748. [CrossRef]

19. Górecki, M.; Jabłońska, E.; Kruszewska, A.; Suszczyńska, A.; Urbańczyklipkowska, Z.; Gerards, M.; Morzycki, J.W.; Szczepek, W.J.; Frelek, J. Practical method for the absolute configuration assignment of tert/tert 1,2-diols using their complexes with $Mo_2(OAc)_4$. *J. Org. Chem.* **2007**, *72*, 2906–2916. [CrossRef] [PubMed]

20. Wang, W.J.; Li, D.Y.; Li, Y.C.; Hua, H.M.; Ma, E.L.; Li, Z.L. Caryophyllene sesquiterpenes from the marine-derived fungus *Ascotricha* sp. ZJ-M-5 by the one strain–many compounds strategy. *J. Nat. Prod.* **2014**, *77*, 1367–1371. [CrossRef] [PubMed]

21. Takekawa, H.; Tanaka, K.; Fukushi, E.; Matsuo, K.; Nehira, T.; Hashimoto, M. Roussoellols A and B, tetracyclic fusicoccanes from *Roussoella hysterioides*. *J. Nat. Prod.* **2013**, *76*, 1047–1051. [CrossRef] [PubMed]

22. White, K.N.; Amagata, T.; Oliver, A.G.; Tenney, K.; Wenzel, P.J.; Crews, P. Structure revision of spiroleucettadine, a sponge alkaloid with a bicyclic core meager in H-atoms. *J. Org. Chem.* **2009**, *73*, 8719–8722. [CrossRef] [PubMed]

23. Rychnovsky, S.D. Predicting NMR spectra by computational methods: Structure revision of hexacyclinol. *Org. Lett.* **2006**, *8*, 2895–2898. [CrossRef] [PubMed]

24. Zhan, G.; Zhou, J.; Liu, J.; Huang, J.; Zhang, H.; Liu, R.; Yao, G. Acetylcholinesterase inhibitory alkaloids from the whole plants of *Zephyranthes carinata*. *J. Nat. Prod.* **2017**, *80*, 2462–2471. [CrossRef] [PubMed]

25. Wang, Q.X.; Bao, L.; Yang, X.L.; Liu, D.L.; Guo, H.; Dai, H.Q.; Song, F.H.; Zhang, L.X.; Guo, L.D.; Li, S.J. Ophiobolins P-T, five new cytotoxic and antibacterial sesterterpenes from the endolichenic fungus *Ulocladium* sp. *Fitoterapia* **2013**, *90*, 220–227. [CrossRef] [PubMed]

26. Wei, H.; Itoh, T.; Kinoshita, M.; Nakai, Y.; Kurotaki, M.; Kobayashi, M. Cytotoxic sesterterpenes, 6-epi-ophiobolin G and 6-epi-ophiobolin N, from marine derived fungus *Emericella variecolor* GF10. *Tetrahedron* **2004**, *60*, 6015–6019. [CrossRef]

27. Liu, H.B.; Edrada-Ebel, R.A.; Ebel, R.; Wang, Y.; Schulz, B.; Draeger, S.; Müller, W.E.G.; Wray, V.; Lin, W.H.; Proksch, P. Ophiobolin sesterterpenoids and pyrrolidine alkaloids from the sponge-derived fungus *Aspergillus ustus*. *Helv. Chim. Acta* **2011**, *94*, 623–631. [CrossRef]

28. Zhu, T.; Lu, Z.; Fan, J.; Wang, L.; Zhu, G.; Wang, Y.; Li, X.; Hong, K.; Piyachaturawat, P.; Chairoungdua, A. Ophiobolins from the mangrove fungus *Aspergillus ustus*. *J. Nat. Prod.* **2018**, *81*, 2–9. [CrossRef] [PubMed]

29. Vainio, M.J.; Johnson, M.S. Generating conformer ensembles using a multiobjective genetic algorithm. *J. Chem. Inf. Model.* **2007**, *47*, 2462–2474. [CrossRef] [PubMed]

30. Puranen, J.S.; Vainio, M.J.; Johnson, M.S. Accurate conformation-dependent molecular electrostatic potentials for high-throughput in silico drug discovery. *J. Comput. Chem.* **2010**, *31*, 1722–1732. [CrossRef] [PubMed]

31. Frisch, M.J.; Trucks, G.W.; Schlegel, H.B.; Scuseria, G.E.; Robb, M.A.; Cheeseman, J.R.; Scalmani, G.; Barone, V.; Mennucci, B.; Petersson, G.A.; et al. *Gaussian 09, Revision D.01*, Gaussian, Inc.: Wallingford, CT, USA, 2009.

32. Qiao, Y.; Xu, Q.; Hu, Z.; Li, X.N.; Xiang, M.; Liu, J.; Huang, J.; Zhu, H.; Wang, J.; Luo, Z.; et al. Diterpenoids of the cassane type from *Caesalpinia decapetala*. *J. Nat. Prod.* **2016**, *79*, 3134–3142. [CrossRef] [PubMed]

33. Zhan, G.; Liu, J.; Zhou, J.; Sun, B.; Aisa, H.A.; Yao, G. Amaryllidaceae alkaloids with new framework types from *Zephyranthes candida* as potent acetylcholinesterase inhibitors. *Eur. J. Med. Chem.* **2017**, *127*, 771–780. [CrossRef] [PubMed]

MDPI

St. Alban-Anlage 66

4052 Basel

Switzerland

Tel. +41 61 683 77 34

Fax +41 61 302 89 18

www.mdpi.com

Marine Drugs Editorial Office

E-mail: marinedrugs@mdpi.com

www.mdpi.com/journal/marinedrugs

www.ingramcontent.com/pod-product-compliance
Lightning Source LLC
Chambersburg PA
CBHW051911210326
41597CB00033B/6104